江戸から平成の昆虫研究を支えた東京大学秘蔵コレクション

珠玉の昆虫標本

Ultimate Insect Specimens – Treasured Entomology Collection of The University of Tokyo spanning the Edo to Heisei eras

珠玉の昆虫標本
── 江戸から平成の昆虫研究を支えた東京大学秘蔵コレクション ──

Ultimate Insect Specimens
── Treasured Entomology Collection of The University of Tokyo spanning the
Edo to Heisei eras

東京大学総合研究博物館
── 2018 年特別展図録 ──

Catalogue of Special Exhibition 2018
The University Museum, The University of Tokyo

矢後勝也 編著
須田真一・山崎剛史 共著

Edited by Masaya Yago
Written by Masaya Yago, Shin-ichi Suda & Takeshi Yamasaki

東京大学出版会
University of Tokyo Press

図録情報

珠玉の昆虫標本──江戸から平成の昆虫研究を支えた東京大学秘蔵コレクション──

編　著：矢後勝也（東京大学総合研究博物館 助教）
共　著：須田真一（東京大学総合研究博物館 研究事業協力者／日本トンボ学会 役員）、
　　　　山崎剛史（公益財団法人山階鳥類研究所 自然史研究室長）
撮　影：桶田太一（OTPHOTO 代表／むさしの自然史研究会 講師）、
　　　　伊藤勇人（東京大学総合研究博物館 研究事業協力者／駒場東邦中学校・高等学校 非常勤講師）、
　　　　鈴木知之（昆虫写真家）、手代木求（東京大学総合研究博物館 研究事業協力者）、フォワードストローク
装丁・デザイン：関岡裕之（東京大学総合研究博物館 特任准教授）
印刷・製本：秋田活版印刷株式会社
製作・発行：東京大学総合研究博物館

発　売：東京大学出版会
　　　　〒 153-0041 東京都目黒区駒場 4-5-29

Catalogue Information

Ultimate Insect Specimens – Treasured Entomology Collection of The University of Tokyo spanning the Edo to Heisei eras

Author & scientific editor: Masaya Yago
Co-authors: Shin-ichi Suda, Takeshi Yamasaki
Photography: Taichi Okeda, Hayato Ito, Tomoyuki Suzuki, Motomu Teshirogi, Forward Stroke inc.
Book design & binding: Hiroyuki Sekioka
Printing: Akita Kappan Printing Co., Ltd.
Publisher: The University Museum, The University of Tokyo

Distributed by
University of Tokyo Press
4-5-29 Komaba, Meguro-ku, Tokyo, Japan
URL: http://www.utp.or.jp

Copyright © 2018 by The University Museum, The University of Tokyo
All rights reserved. Published December 31, 2018. Second printing March, 2019
Printed in Japan
ISBN 978-4-13-060145-0

展示情報

2018 年特別展示
珠玉の昆虫標本──江戸から平成の昆虫研究を支えた東京大学秘蔵コレクション──

会　期：2018 年 7 月 14 日（土）〜 2018 年 10 月 14 日（日）
会　場：東京大学総合研究博物館（東京都文京区本郷 7-3-1）
主　催：東京大学総合研究博物館
後　援：（公財）山階鳥類研究所、日本昆虫学会、日本鱗翅学会、日本蝶類科学学会、日本蝶類学会
企　画・総指揮：矢後勝也
副指揮：須田真一・谷尾 崇
展示デザイン：洪 恒夫
グラフィックデザイン：関岡裕之・西野瞳子

Exhibit Information

Special Exhibition 2018
Ultimate Insect Specimens –Treasured Entomology Collection of The University of Tokyo spanning the Edo to Heisei eras

Exhibition period: July 14 - October 20, 2018
Exhibition hall: The University Museum, The University of Tokyo
Host: The University Museum, The University of Tokyo
Supporting organizations: Yamashina Institute for Ornithology, The Entomological Society of Japan,
　The Lepidopterological Society of Japan, The Butterfly Science Society of Japan, The Butterfly Society of Japan
Planner and Executive Producer: Masaya Yago
Vice-Producers: Shin-ichi Suda and Takashi Tanio
Exhibition Design: Tsuneo Ko
Graphic Design: Hiroyuki Sekioka and Toko Nishino

目次

序文		009
ごあいさつ	諏訪　元	010
はじめに	矢後　勝也	012
第1章（江戸〜昭和初期［戦前］）		015
武蔵石寿 ── 日本最古の昆虫標本の作製者	矢後　勝也	016
佐々木忠次郎 ── 近代養蚕学・農業害虫学の開祖	矢後　勝也	020
箕作佳吉 ── 日本人最初の動物学教授	矢後　勝也	025
加藤正世 ── 昆虫黄金期を築いたセミ博士	矢後　勝也	030
山階芳麿 ── 山階鳥類研究所の創立者	山崎　剛史	046
第2章（昭和中期［戦後］〜平成）		059
五十嵐邁 ── 蝶類幼生期研究の大家	矢後　勝也	060
江田　茂 ── 国内有数の大収集家	矢後　勝也	074
須田孫七 ── 日本のファーブル	須田　真一	083
濱　正彦 ── 信州の蝶聖	矢後　勝也	096
宮野浩二 ── 西多摩の博物学者	矢後　勝也	105
白石浩次郎 ── 平和・トンボ資料館館長	須田　真一	117
石川良輔 ── ハチ・オサムシ研究の巨匠	矢後　勝也	121
尾本惠市 ── 二刀流の東京大学名誉教授	矢後　勝也	130
岸田泰則 ── 日本蛾類学会会長	矢後　勝也	139
第3章		151
幻の大蝶「ブータンシボリアゲハ」 　　　　── ブータン国王陛下からの贈呈標本	矢後　勝也	152
昆虫 ── 東京大学総合研究博物館データベース	矢後　勝也	156
未来に向けて	矢後　勝也	158
第4章		159
展示制作 ── 立案から完成まで	矢後　勝也	160
会場風景		162
参考文献		165

Contents

Preface		009
Greeting	Gen Suwa	010
Outline	Masaya Yago	012
Chapter 1 (Edo to early Showa eras)		015
Sekiju Musashi — Creator of the oldest insect specimens in Japan	Masaya Yago	016
Chûjirô Sasaki — Founder of modern sericulture and pestology in Japan	Masaya Yago	020
Kakichi Mitsukuri — The first professor of zoology in Japan	Masaya Yago	025
Masayo Kato — "Doctor Cicada", creator of the Golden Age of Insects	Masaya Yago	030
Yoshimaro Yamashina — Founder of the Yamashina Institute for Ornithology	Takeshi Yamasaki	046
Chapter 2 (Mid-Showa to Heisei eras)		059
Suguru Igarashi — An authority on the early stages of butterflies	Masaya Yago	060
Shigeru Eda — One of the greatest insect collectors in Japan	Masaya Yago	074
Magoshichi Suda — Regarded as the "Japanese Fabre"	Shin-ichi Suda	083
Masahiko Hama — Saint Aurelian of Shinshu	Masaya Yago	096
Koji Miyano — Great naturalist from Nishitama, Tokyo	Masaya Yago	105
Kojiro Shiraishi — Director of the Peace and Dragonfly Museum	Shin-ichi Suda	117
Ryôsuke Ishikawa — Great entomologist studying bees and carabid beetles	Masaya Yago	121
Keiichi Omoto — Professor emeritus of the University of Tokyo, a leading anthropologist and lepidopterist	Masaya Yago	130
Yasunori Kishida — President of the Japan Heterocerists' Society	Masaya Yago	139
Chapter 3		151
Ludlow's Bhutan Glory, a mysterious swallowtail — Butterfly specimen presented by His Majesty the King of Bhutan	Masaya Yago	152
INSECTS — The University Museum Database, the University of Tokyo (UMDB)	Masaya Yago	156
Looking towards the future	Masaya Yago	158
Chapter 4		159
Exhibition production — From plan to completion	Masaya Yago	160
Exhibition views		162
References		165

序文

Preface

ごあいさつ

　東京大学総合研究博物館は、前身の総合研究資料館発足から数えて 52 年目を迎えました。創学以来、集積された多くの貴重な学術標本コレクションの管理・運営・継承の重責を担いながら、コレクションの一層の充実、さらには最新の研究とその公開発信に邁進しています。

　本館ではこれまでに多くの特別展示を開催してきましたが、昆虫の特別展は初めての試みとなります。蓄積された昆虫標本群は、歴史ある本学の昆虫標本から近年の寄贈によるコレクションまでさまざまです。今回は、主に 15 名の研究者により収集された標本について、概説を交えながら展示しました。それぞれの研究者が抱いた昆虫に寄せる興味と学術的探究への情熱が、これらの昆虫標本から鮮烈に感じ取れることでしょう。また、昆虫は環境を強く反映する生き物です。そのため、これらの昆虫標本から過去の昆虫相の変遷を調査することで、近年の環境変化を直接的に読み取ることもできます。当館では、文化現象も含む自然界のあり方とその地球史的経緯について理解を深めることを目標に掲げていますが、まさにこの理念を踏襲する学術知を昆虫標本からも生み出すことが可能と言えます。

　色・形の多種多様な昆虫の壮麗さを全空間に表し、圧倒的な迫力を引き出すことを目指した本展示により、「標本・資料」が持つ研究世界や自然への関心を深める機会となれば幸いです。この特別展の開催にあたり、ご協力頂いた関係諸機関、諸氏に厚くお礼申し上げます。

諏訪 元 (東京大学総合研究博物館 館長)

Greetings

We welcome you to our newest special exhibit at The University Museum, The University of Tokyo (UMUT), our first exhibit specifically featuring the UMUT entomology collections and related research. In this exhibit, we show for the first time a grand time series of our insect collections, representing the historical collections of the University of Tokyo to more recently donated specimens. We feature the collections of 15 researchers, which show their irresistible affinities to these specimens and their devotion to scientific enquiry. As a zoological reality, insects keenly reflect their environments. This enables the documentation and evaluation of past and present environmental parameters, some such aspects that we here show. Above all, we welcome you to enjoy the sheer variety and diversity that the entomological world has to offer, that we show and extend here in this exhibit hall with all our might.

Gen Suwa

Director / The University Museum, The University of Tokyo

はじめに

　日本の昆虫学は東京大学に端を発し、様々な学術分野や研究機関に枝分かれして今に至ります。この学問の発展には専門機関の研究者だけでなく、むしろ在野の研究者の貢献も大きいところです。その間、学術研究や教育普及のために収集され、本学に集積・寄贈されてきた昆虫標本も膨大な数に及びます。

　本特別展では、東京大学総合研究博物館に収蔵されている約70万点の昆虫標本のうち、日本の昆虫研究史の源流ともいえる学術標本から現在に至るまで継続的に収集、研究されてきた秘蔵コレクション約55,711点を一挙公開しました。

　この中には約200年前の江戸時代に生きた本草学者・武蔵石寿の作製による日本最古の昆虫標本、近代養蚕学の父・佐々木忠次郎やミツクリザメで知られる箕作佳吉の明治〜大正期の昆虫標本、昭和初期に採集された鳥類学者の侯爵・山階芳麿やセミ博士と呼ばれた加藤正世の昆虫標本、ブータン国王陛下から贈呈されたブータンシボリアゲハ、昆虫学史上に名を連ねる五十嵐邁、江田 茂、須田孫七、濱 正彦、宮野浩二、白石浩次郎、石川良輔、尾本惠市、岸田泰則などのコレクションが含まれます。

　これらの自然史遺産ともいえる貴重な昆虫標本を一堂に展示することで、いわば日本の昆虫博物誌を体感してもらうことを一つの趣意としています。また、これを機に多様な昆虫への幅広い興味や科学的な探究心を抱いてもらえたら幸いです。

　この特別展示の開催および図録の出版にあたり、下記の方々にご支援、ご協力頂きました。(公財) 山階鳥類研究所、日本昆虫学会、日本鱗翅学会、日本蝶類科学学会、日本蝶類学会、東京大学大学院農学生命科学研究科応用昆虫学研究室・昆虫遺伝研究室、ミキモト真珠島、五十嵐昌子、石川良輔、伊藤勇人、井上暁生、上島 励、大野正男、桶田太一、尾本惠市、柿沼 隆、勝山礼一朗、岸田泰則、久保田繁男、斎藤基樹、清水 晃、瀬戸山知佳、世良裕朝、添徹太郎、高木大司、武智昭一、谷尾 崇、築山 洋、津久井岳、手代木求、長瀬博彦、新津修平、原田基弘、堀江洋成、前川 優、Neil Moffat、矢野高広、杠 隆史、吉田良和、東京大学総合研究博物館ボランティアの方々 (敬称略・順不同)。この場を借りて、心よりお礼を申し上げます。

　　　　　　　　　　　　　　　　　　　　　　　　矢後 勝也 (東京大学総合研究博物館 助教)

Exhibition Outline

The roots of entomology in Japan were established at the University of Tokyo, where it has grown and branched into the numerous academic fields and research institutes found around Japan today.

This development has been made thanks to a passionate community of not only professional researchers but also keen amateurs. During the history of Japanese entomology, this community has provided our university museum with a wealth of insect specimens for academic research and educational dissemination.

We have drawn from our vast collection of over 700,000 insect specimens to show in this special exhibition our most treasured, comprised of 55,711 specimens including the oldest insect specimens in Japan, made by Sekiju Musashi in the Edo era (about 200 years ago). This collection also contains old specimens dating back to the Meiji and Taisho eras (1868 – 1926), collected by our university's Prof. Chûjirô Sasaki and Prof. Kakichi Mitsukuri as well as insect specimens collected by Dr. Yoshimaro Yamashina and Dr. Masayo Kato in the early Showa era. The exhibition also gives you the rare opportunity to see the specimen of a recently rediscovered butterfly (Bhutanitis ludlowi) donated by the King of Bhutan, and the precious insect specimens of Dr. Suguru Igarashi, Mr. Shigeru Eda, Mr. Magoshichi Suda, Mr. Masahiko Hama, Mr. Koji Miyano, Mr. Kojiro Shiraiwa, Prof. Emer. Ryôsuke Ishikawa, Prof. Emer. Keiichi Omoto and Mr. Yasunori Kishida who made great achievements in the history of entomology.

The purpose of this special exhibition is to communicate the natural history with Japanese entomology from the significant heritage of the insect specimens preserved in our museum. Moreover, we hope visitors develop a greater understanding and scientific curiosity for the diverse insects of not only this collection but of all entomology.

In the special exhibition, we would like to express our great thanks to Yamashina Institute for Ornithology, The Entomological Society of Japan, The Lepidopterological Society of Japan, The Butterfly Science Society of Japan, The Butterfly Society of Japan, Laboratory of Applied Entomology and Laboratory of Insect Genomics and Bioscience, Graduate School of Agricultural and Life Sciences, the University of Tokyo, Mikimoto Pearl Island, Yoshiko Igarashi, Ryôsuke Ishikawa, Hayato Ito, Akeo Inoue, Rei Ueshima, Masao Ohno, Taichi Okeda, Keiichi Omoto, Takashi Kakinuma, Raiichiro Katsuyama, Yasunori Kishida, Shigeo Kubota, Motoki Saito, Akira Shimizu, Chika Setoyama, Hirotomo Sera, Tetsutaro Soe, Taishi Takagi, Shoichi Takechi, Takashi Tanio, Hiroshi Tsukiyama, Gaku Tsukui, Motomu Teshirogi, Hirohiko Nagase, Shuhei Niitsu, Motohiro Harada, Hironari Horie, Yu Maekawa, Takeshi Yamasaki, Takahiro Yano, Takashi Yuzuriha, Yoshikazu Yoshida and Neil Moffat for their kind support and valuable advice.

Masaya Yago

Assistant Professor / The University Museum, The University of Tokyo

第 1 章

江戸〜昭和初期［戦前］

Chapter 1
Edo to early Showa eras

武蔵石寿
——日本最古の昆虫標本の作製者

Sekiju Musashi
— Creator of the oldest insect specimens in Japan

武蔵石寿の標本箱の蓋

江戸時代の旗本・武蔵石寿（1766～1861年）は江戸（東京）生まれの本草学者で、富山藩主・前田利保が主宰する博物研究会「赭鞭会」の主要メンバーの一人である。江戸時代の動植物学誌で最高峰の貝類図譜「目八譜」の著者でもある。江戸時代末期（天保年間：1830～1844年）に作られた武蔵石寿昆虫コレクションは、国内最古の昆虫標本として知られる。

この歴史的価値の高い昆虫コレクションは長く行方が分からなかったが、1913年に仏国大使館の外交官エドム・アンリ・ガロア（1878～1956年）が東京の古道具屋で見出した。当初は東京の帝室博物館（現・東京国立博物館）に寄贈しようとしたが、出願書の提出などの面倒な手続きに憤慨し、東京帝国大学農科大学（現・東大農学部）の佐々木忠次郎教授（1857～1938）に寄贈した。ガロアは有名な昆虫愛好家でもあり、在日中に日光でガロアムシ目ガロアムシ科のガロアムシを発見したことから、彼の名が属の学名および標準和名として付けられている。

石寿の標本は上部がドーム型のガラス容器に覆われている。中には綿が敷かれ、その上に虫が置かれていて、底面には厚い和紙が貼り付けられている。この虫針を使わずにガラス容器に封じられた独特の作製法は、世界でも他に類を見ない。ガラス容器は通常のガラスよりも鉛の含有量が高い。鉛ガラスの利点としては、屈折率が高く輝きが増すことの他、融点が低いために低温でも加工しやすいという。

これらの中には、アオスジアゲハやハンミョウ、ギンヤンマなど9目約72種の昆虫が含まれる。タガメやナミゲンゴロウ、ゴミアシナガサシガメのような絶滅危惧種もいる。その他にカニ、アブラコウモリ、トカゲなども見られるが、その多くは漢字で書くと虫偏が付くことから、当時認識されていた昆虫の範囲の広さが分かる貴重な資料である。2012年に農学部から総合研究博物館に移管されて、現在に至っている。

（矢後勝也）

Mr. Sekiju Musashi (1766–1861), a direct retainer of the shogun in the Edo era, was a herbalist born in Tokyo (then known as Edo) and one of the main members of the Shaben-kai, a society for natural history, which Toshiyasu Maeda (the lord of the Toyama Domain) presided over. He was also the author of Mokuhachifu, the most respected illustrated book of shells from the Edo era. His insect collection, made in the late Edo era (the Tenpo era: 1830–1844), is known for being comprised of the oldest insect specimens in Japan.

The historically-valuable collection was lost after his death until 1913, when Edme Henri Gallois (1878–1956), a resident of the Embassy of France, stumbled upon it at a second-hand store in Tokyo. At first he attempted to give it to the Imperial Museum, Tokyo (now the Tokyo National Museum). However, after becoming angered at the Imperial Museum's demands for official contracts and extensive paperwork, he instead donated it to Prof. Chûjirô Sasaki (1857–1938) in the College of Agriculture, Imperial University of Tokyo (now the Faculty of Agriculture, The University of Tokyo). Gallois was also himself a famous amateur entomologist, discovering a new species, "Garoamushi" *Galloisiana nipponensis* (Grylloblattodea: Grylloblattidae), with the generic and Japanese names named after him.

The specimens in the collection are notable for their unique containers, which resemble snow globes. The specimens are covered with dome-shaped glass containers, in which insects are placed on large balls of cotton wool sitting atop discs of thick traditional Japanese paper. This unique preparation method, holding insects in glass containers without pins, is unprecedented in the world. The glass containers are made of crystal glass, which is much higher in lead content than modern glassware. Two advantages of the lead glass are that it has a stronger brilliance due to a high refractive index, and that it is easy to process at lower temperature due to its low melting point.

The insect collection contains 72 insect species represented as nine orders including butterflies, tiger beetles and dragonflies. In addition, it includes a bat, crabs and lizards, most of which are notable for having the name "mushi-hen" (represented by the radical for "insect" in their Chinese character names). Thus, these specimens are precious in their ability to make us understand the wide range of insects at that time.

This collection was transferred from the Faculty of Agriculture to the University Museum in 2012.

(Masaya Yago)

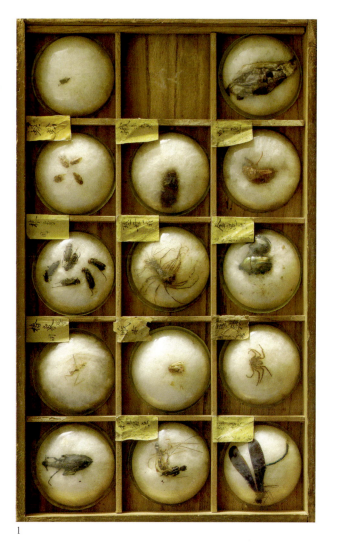

1. ハグロトンボ［カワトンボ科］（蜻蛉目）・ゴミアシナガサシガメ［サシガメ科］（半翅目）・マイマイカブリ［オサムシ科］（鞘翅目）他.
Atrocalopteryx atrata [Calopterygidae] (Odonata), *Myiophanes tipulina* [Reduviidae] (Hemiptera), *Damaster blaptoides* [Carabidae] (Coleoptera), etc.

2

3

4

2. クツワムシ[キリギリス科](直翅目)・ミンミンゼミ[セミ科](半翅目)・アブラコウモリ[ヒナコウモリ科](蝙蝠目[翼手目])他.
 Mecopoda nipponensis [Tettigoniidae] (Orthoptera), *Hyalessa maculaticollis* [Cicadidae] (Hemiptera), *Pipistrellus abramus* [Vespertilionidae] (Chiroptera), etc.

3. ケラ[ケラ科](直翅目)・ナミゲンゴロウ[ゲンゴロウ科](半翅目)・ジョロウグモ[ジョロウグモ科](蜘蛛目)他.
 Gryllotalpa orientalis [Gryllotalpidae] (Orthoptera), *Cybister japonicus* [Dytiscidae] (Coleoptera), *Nephila clavata* [Nephilidae] (Araneae), etc.

4. ハンミョウ[オサムシ科]・ナミハナムグリ[コガネムシ科]・ヒゲコメツキ[コメツキムシ科](鞘翅目)他.
 Cicindela japonica [Carabidae], *Cetonia pilifera* [Scarabaeidae], *Pectocera fortunei* [Elateridae] (Coleoptera), etc.

5

6

7

5. タイコウチ［タイコウチ科］（半翅目）・クスサン繭［ヤママユ科］（鱗翅目）・ヤマトタマムシ［タマムシ科］（鞘翅目）他．
 Laccotrephes japonensis [Nepidae] (Hemiptera), cocoon of *Caligula japonica* [Saturniidae] (Lepidoptera), *Chrysochroa fulgidissima* [Buprestidae] (Coleoptera), etc.

6. ナミアゲハ蛹［アゲハチョウ科］・ヤママユ繭［ヤママユ科］（鱗翅目）・ヒダリマキマイマイ［マイマイ科］（有肺目）他．
 Pupa of *Papilio xuthus* [Papilionidae], cocoon of *Antheraea yamamai* [Saturniidae] (Lepidoptera), *Euhadra quaesita* [Bradybaenidae] (Pulmonata), etc.

7. ギンヤンマ［ヤンマ科］（蜻蛉目）・タガメ［コオイムシ科］（半翅目）・アオスジアゲハ［アゲハチョウ科］（鱗翅目）他．
 Anax parthenope [Aeshnidae] (Odonata), *Lethocerus deyrollei* [Belostomatidae] (Hemiptera), *Graphium sarpedon* [Papilionidae] (Lepidoptera), etc.

1-7. 箱寸法 (Box dimantions) 237×390 mm.

019

佐々木忠次郎
——近代養蚕学・農業害虫学の開祖

Chûjirô Sasaki
— Founder of modern sericulture and pestology in Japan

佐々木忠次郎（1857〜1938）

明治〜昭和初期に活躍した佐々木忠次郎（1857〜1938年）は、帝国大学（のちの東京帝国大学）農科大学養蚕学教室の初代教授の一人である。日本の近代養蚕学や農業害虫学の開祖として知られ、応用昆虫学の礎を築き、日本初の昆虫学を講じた。特にカイコの病虫害の研究とその対策により日本の養蚕業に大きく貢献した。当時の研究の様子は、図示したカイコの標本や害虫の標本からもよく分かる。これらの研究の功績が称えられ、没後に旭日重光章が贈られている。国蝶オオムラサキの属名 *Sasakia* が献名されていることでも有名である。一方、東京大学の学生時に「大森貝塚」の発見者エドワード・S・モースの指導を受け、その貝塚の発掘調査にも携わった。また、茨城県で「陸平貝塚」を発見し、飯島魁と日本人のみで最初の発掘調査を行うなど、考古学分野にも名を残している。

佐々木コレクションは本学農学部から2012年に移管されたもので、欧米式の針刺し標本としては国内最古級となる。明治〜大正期の昆虫標本が主で、佐々木教授の他、三宅恒方、長野菊次郎、石原保など、著名な昆虫学者由来の昆虫標本や彼らが記載した新名のタイプ標本が含まれている他、東京産を中心とした絶滅産地の昆虫も多く見られる。

（矢後勝也）

Dr. Chûjirô Sasaki (1857–1938), active in the Meiji to early Showa eras, was one of the first professors at the Laboratories of Sericulture, College of Agriculture, Imperial University (later Tokyo Imperial University). He is known as the founder of modern sericulture and pestology in Japan, notably being the first person in Japan to give lectures on entomology. In particular, he contributed to the development of sericulture industry in Japan due to studies on insect pests and diseases of silkworms. The illustrated specimen boxes of silkworms and harmful insects show the research situation of that time in his laboratory. In recognition of this contribution, he was posthumously awarded "The Order of the Rising Sun, Gold and Silver Star" from the government of Japan. Furthermore, the generic name of the National Butterfly of Japan, *Sasakia charonda*, was named after him. In his university days, Dr. Sasaki took lectures from Prof. Edward S. Morse who discovered the Ohmori Shell Mounds, which Dr. Sasaki helped to excavate. Moreover, Dr. Sasaki and Dr. Isao Iijima found the Okadaira Shell Mound in Ibaraki, a major contribution to Japanese archaeology.

The Sasaki and Associated Researchers' Collection, which was transferred from the Faculty of Agriculture to the University Museum in 2012, is one of the oldest collections of European-style pinned specimens in Japan. Moreover, his collection includes many specimens of endangered species that are now extinct in Tokyo and other localities, collected by prominent entomologists such as Dr. Tsunekata Miyake, Mr. Kikujiro Nagano and Dr. Tamotsu Ishihara. The collection also contains the holotype specimen of new taxa which they had described.

(Masaya Yago)

1. カイコ［カイコガ科］（鱗翅目）.
   Silkmoth, *Bombyx mori* [Bombycidae] (Lepidoptera).

2. 野菜の害虫.
   Vegetable insect pests.

3. 大豆の害虫.
   Soybean insect pests.

4. 東京産ベッコウトンボ・ヨツボシトンボ他［トンボ科］（蜻蛉目）.
   Libellula angelina & *L. quadrimaculata* from Tokyo, etc. [Libellulidae] (Odonata).

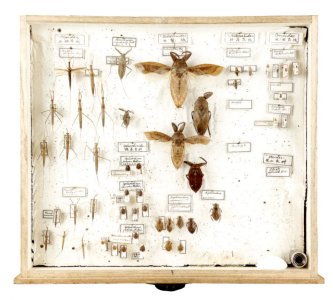

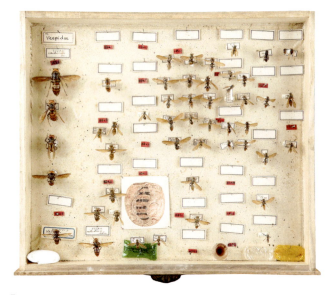

5. タガメ［コオイムシ科］・ヒメミズカマキリ［タイコウチ科］・トゲナベブタムシ ［ナベブタムシ科］他（半翅目）.
   Lethocerus deyrollei [Belostomatidae], *Ranatra chinensis* [Nepidae], *Aphelocheirus nawai* [Aphelocheiridae], etc. (Hemiptera).

6. 東京産スジゲンゴロウ・コガタノゲンゴロウ他［ゲンゴロウ科］（鞘翅目）.
   Hydaticus satoi & *Cybister tripunctatus* from Tokyo, etc. [Dytiscidae] (Coleoptera).

7. オオスズメバチ・モンスズメバチ他［スズメバチ科］（膜翅目）.
   Vespa mandarinia, *V. crabro*, etc. [Vespidae] (Hemiptera).

8

9

8. 東京産オオウラギンヒョウモン・ウラギンヒョウモン他［タテハチョウ科］（鱗翅目）.
   Fabriciana nerippe & *F. adippe* from Tokyo, etc. [Nymphalidae] (Lepidoptera).

9. オオムラサキ・コムラサキ他［タテハチョウ科］（鱗翅目）.
   Sasakia charonda, Apatura metis, etc. [Nymphalidae] (Lepidoptera).

10

11

10. 東京産チャマダラセセリ・ミヤマセセリ・ダイミョウセセリ他［セセリチョウ科］（鱗翅目）.
Pyrgus maculatus, Erynnis montanus & *Daimio tethys* from Tokyo, etc. [Hesperiidae] (Lepidoptera).

11. ムネアカセンチコガネ科・センチコガネ科・コガネムシ科（鞘翅目）.
Bolboceratidae, Geotrupidae & Scarabaeidae (Coleoptera).

1-11. 箱寸法 (Box dimantions) 363×435 mm.

箕作佳吉
―― 日本人最初の動物学教授

Kakichi Mitsukuri
— The first professor of zoology in Japan

箕作佳吉（1858〜1909）

箕作佳吉（1858〜1909年）は、日本人最初の動物学教授として知られる。江戸（東京）で生まれ、留学先のイェール大学やジョンズ・ホプキンズ大学、ケンブリッジ大学などで動物学を学んだ。帰国後、1882年に東京大学理学部で日本人最初の動物学教授となる。1883年にジョンズ・ホプキンズ大学から学術博士、1888年に帝国大学（のちの東京帝国大学）から理学博士を授与され、1901〜1907年に東京帝国大学理科大学長も務めた。箕作教授は日本の動物学創成期の指導者であったため、献名された和名や学名が多く、特にミツクリザメは有名で、昆虫でもミツクリハバチ、ミツクリクロタマゴバチ、ミツクリヒゲナガハナバチなどが知られる。

箕作昆虫コレクションには、箕作教授の講義用標本や著名な昆虫家の採集品など、歴史的価値は高いものがある。特に、ギフチョウの発見者として知られる名和 靖氏由来の標本が多い。貴族院議員だった高千穂宣麿（1865〜1950年）由来のエルタテハやコヒョウモンモドキ、旧制第五高等学校（現・熊本大学）の教授を務めた中川久知（1859〜1921年）由来の熊本産ミカドアゲハ、のちにハノーヴァーの州立博物館長や動物園長となったドイツの著名な動物学者アドルフ・フリッツェ（1860〜1927年）が採集したオオキノコムシなども含まれる。また、東京産のマルコガタノゲンゴロウやナミゲンゴロウ、カワラハンミョウ、オオクワガタのような絶滅産地の昆虫標本なども見られる。本館資料部動物部門の所蔵品。

（矢後勝也）

Dr. Kakichi Mitsukuri (1858–1909) was the first professor of zoology in Japan. He studied zoology at Yale University, Johns Hopkins University and the University of Cambridge. After his return, he became the first professor of zoology in Japan presiding at the College of Science, Tokyo University. In 1883, he received a Ph.D. from Johns Hopkins University and in 1888 a doctorate of science from Imperial University (later Tokyo Imperial University) and in 1901–1907 held the position of President of the College. Since Prof. Mitsukuri was a prominent leader in the founding period of zoology in Japan, the Japanese and Scientific names of many animals and insects were named in his honour. Examples of this are the shark *Mitsukurina owstoni* Jordan, 1898, and the bees, *Eriocampa mitsukurii* Rohwer, 1910, *Trissolcus mitsukurii* (Ashmead, 1904) and *Tetralonia mitsukurii* Cockerell, 1911.

His insect collection contains numerous historically valuable specimens including those used in his lectures and others collected by famous entomologists. In particular, there are many collected by Yasushi Nawa, the entomologist who discovered the papilionid butterfly *Luehdorfia japonica*. The collection also contains specimens of nymphalid butterflies *Nymphalis vaualbum* and *Melitaea ambigua* collected by a member of the House of Lords, Mr. Nobumaro Takachiho (1865–1950). A specimen of a papilionid *Graphium doson*, collected by a teacher at the former Fifth High School (now Kumamoto University), Prof. Hisatomo Nakagawa (1859–1921), is also part of the collection, as are specimens of a fungus beetle *Encaustes praenobilis*, collected by a famous German zoologist, Dr. Adolf Fritze (1860–1927), who became Director of the State Museum and President of the

Zoological Garden in Hannover. Moreover, this collection includes specimens of many species faced with extinction from the Tokyo area such as *Cybister japonicus*, *Cybister lewisianus*, *Cicindela laetescripta* and *Dorcus hopei*.

(Masaya Yago)

1

2

3

1. コヒョウモンモドキ・メスグロヒョウモン他［タテハチョウ科］（鞘翅目）.
   Melitaea ambigua, *Damora sagana*, etc. [Nymphalidae] (Lepidoptera).

2. アカタテハ・リュウキュウムラサキ他［タテハチョウ科］（鞘翅目）.
   Vanessa indica, *Hypolimnas bolina*, etc. [Nymphalidae] (Lepidoptera).

3. 東京産カワラハンミョウ・ハンミョウ［ハンミョウ科］・アオオサムシ［オサムシ科］他（鞘翅目）.
   Cicindela laetescripta & *C. japonica* from Tokyo [Cicindelidae], *Carabus insulicola* [Carabidae], etc. (Coleoptera).

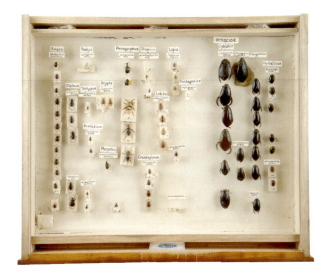

4

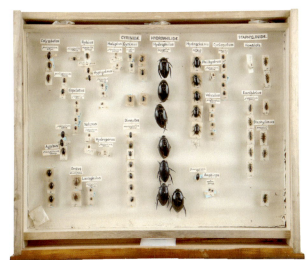

5

6

4. 東京産マルコガタノゲンゴロウ・ゲンゴロウ［ゲンゴロウ科］・ミイデラゴミムシ
   ［オサムシ科］他（鞘翅目）.
   Cybister lewisianus & *C. japonicus* from Tokyo [Dytiscidae], *Pheropsophus jessoensis*
   [Carabidae], etc. (Coleoptera).

5. 東京産ガムシ［ガムシ科］・オオミズスマシ［ミズスマシ科］・ハイイロハネカ
   クシ［ハネカクシ科］他（鞘翅目）.
   Hydrophilus acuminatus [Hydrophilidae], *Dineutus orientalis* [Gyrinidae] &
   Eucibdelus japonicus [Staphylinidae] from Tokyo, etc. (Coleoptera).

6. 東京産オオクワガタ・ヒラタクワガタ［クワガタムシ科］・ダイコクコガネ・ミヤ
   マダイコクコガネ［コガネムシ科］他（鞘翅目）.
   Dorcus hopei & *D. titanus* from Tokyo [Lucanidae], *Copris ochus* & *C. pecuarius*
   [Scarabaeidae], etc. (Coleoptera).

7

8

7. カブトムシ・コカブトムシ・ヤマトアオドウガネ他[コガネムシ科](鞘翅目).
Allomyrina dichotoma, Eophileurus chinensis, Anomala japonica, etc. [Scarabaeidae] (Coleoptera).

8. オオセンチコガネ[センチコガネ科]・シロスジコガネ[コガネムシ科]他(鞘翅目).
Phelotrupes auratus [Geotrupidae], *Polyphylla albolineata* [Scarabaeidae], etc. (Coleoptera).

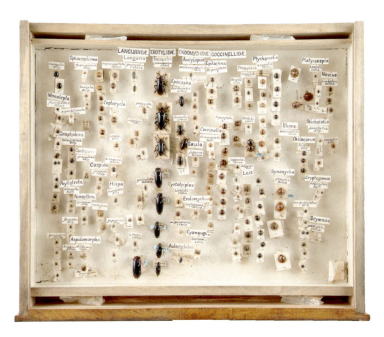

9

10

9. オオテントウ[テントウムシ科]・オオキノコムシ[オオキノコムシ科]・クロトゲ
ハムシ[ハムシ科]他(鞘翅目).
  Synonycha grandis [Coccinellidae], *Encaustes praenobilis* [Erotylidae], *Hispellinus moerens* [Chrysomelidae], etc. (Coleoptera).

10. アカガネカミキリ・コブヤハズカミキリ他[カミキリムシ科](鞘翅目).
  Plectrura metallica, Mesechthistatus binodosus, etc. [Cerambycidae] (Coleoptera).

1-10. 箱寸法(Box dimantions) 392×460 mm.

加藤正世
——昆虫黄金期を築いたセミ博士

Masayo Kato
— "Doctor Cicada", creator of the Golden Age of Insects

加藤正世（1898〜1967）

栃木県生まれの加藤正世博士は、大正から昭和初期に活躍した稀代の昆虫学者である。東京・石神井に「蟬類博物館」を開館し、展示を通じた昆虫学の普及に尽力した。趣味の昆虫採集を通じた教育普及活動にも力を注ぎ、たくさんの少年・青年たちに影響を与え、昭和初期の昆虫黄金時代を築き上げた主要な人物としても知られる。特にセミ・ツノゼミに造詣が深く、多くの新種・新亜種を記載したり、謎の生態を次々と解明したりして、「セミ博士」の愛称としても親しまれていた。

1916年に東京の攻玉社中学校を卒業後、国民飛行会（帝国飛行協会）の記者となり、1922年には三等飛行機操縦士の免許を取得するなど、飛行機に関する造詣にも深いことから、複数の航空関係の書籍も出版している。1956年には大著「蟬の生物学」を発行し、1958年に北海道大学から理学博士の学位を授与された。1962年に長年の自然教育への貢献から藍綬褒章を授与されている。

加藤コレクションは、昆虫だけに留まらず、あらゆる生物を網羅し、その種数も約6万点と膨大である。また、自身の新名記載による約400頭のホロタイプ標本も含まれ、その多くがセミ・ツノゼミ類などの半翅目である。ジェネラル・コレクションには、1930年代の石神井を中心とした東京近郊で採集されたものが多く、当時の関東平野の生物相を知る上で非常に重要な資料と言える。また、1923〜1928年の台湾および1928〜1930年の京都衣笠に住んでいた頃の標本も多い。2010年11月に当時の管理者であった加藤博士の五女・鈴木蘭子氏から東京大学総合研究博物館に寄贈された。

（矢後勝也）

Dr. Masayo Kato (1898 – 1967), born in the region of Tochigi, was a remarkable entomologist active in the 1920s-1960s. In 1938, he opened a private museum, the "Cicadidae Museum" in Shakujii-koen, Tokyo, which he used to enhance and enrich entomological education through exhibitions of insect specimens. His devotion inspired all generations, causing a surge in the popularity of entomology, especially through insect collecting. As a result, he is also known as a key person in creating "the Golden Age of Insects" in the early Showa era. In particular, he had a detailed knowledge of cicadas and treehoppers; not only describing many new taxa of the two groups, but also revealing a lot of their previously unknown ecology. This earned him the nickname "Semi Hakase (Doctor Cicada)".

After graduating from Kogyokusha Junior High School, Tokyo in 1916, he started working as a reporter for "Nihon-hiko-kai". In 1920, while still working there, he became a trainee pilot of a flight institute "Ito-Hikoki-Kenkyujyo" in Tsudanuma, Chiba. Since he was also profoundly knowledgeable about airplanes, he also published several books and articles on aircraft. In 1956, Kato published a major book "The Biology of the Cicadas", which became his PhD dissertation, accepted by Hokkaido University in 1958. In recognition of his contribution, he eventually received the "Medal with Blue Ribbon" from the government of Japan in 1962.

The Kato collection contains not only insects but also other animals and plants, consisting of approximately 60,000 specimens. Moreover, his collection also includes the holotype of approximately 400 new taxa described by himself. Most of the new taxa are Homoptera such as cicadas and grasshoppers. In his general collection, there are many specimens from the 1930s, collected in Tokyo and its surroundings including Shakujii, the location of his museum. These specimens are very important in understanding the insect fauna in the Kanto Plain at that time. Moreover, there are also many specimens from Taiwan and Kyoto, where he lived in 1923-1928 and 1928-1930, respectively. The donation of his collection to the University Museum, The University of Tokyo was carried out according to the wishes of his fifth daughter, Mrs. Sonoko Suzuki, in 2010.

(Masaya Yago)

1

2

1. セミ科・ツノゼミ科・アワフキムシ科他（半翅目）.
   Cicadidae, Membracidae, Aphrophoridae, etc. (Hemiptera).

2. ニイニイゼミ属［セミ科］（半翅目）.
   Platypleura spp. [Cicadidae] (Hemiptera).

3

4

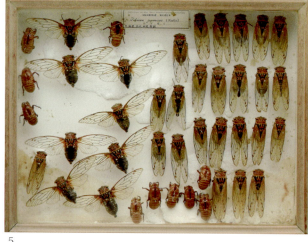
5

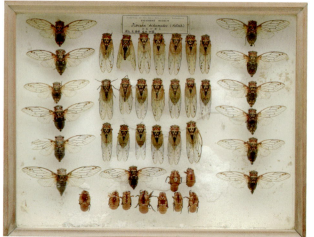
6

3. ニイニイゼミ属 [セミ科] (半翅目).
   Platypleura spp. [Cicadidae] (Hemiptera).

4. ハゴロモゼミ [セミ科] (半翅目).
   Chremistica ochracea [Cicadidae] (Hemiptera).

5. エゾゼミ [セミ科] (半翅目).
   Lyristes japonicus [Cicadidae] (Hemiptera).

6. コエゾゼミ [セミ科] (半翅目).
   Lyristes bihamatus [Cicadidae] (Hemiptera).

7

9

8

10

7. エゾゼミ属 [セミ科] (半翅目).
   Lyristes spp. [Cicadidae] (Hemiptera).

8. クマゼミ・スジアカクマゼミ (チョウセンクマゼミ) [セミ科] (半翅目).
   Cryptotympana facialis & *C. atrata* [Cicadidae] (Hemiptera).

9. リュウキュウクマゼミ・タイワンクマゼミ [セミ科] (半翅目).
   Cryptotympana facialis okinawana & *C. holsti* [Cicadidae] (Hemiptera).

10. クマゼミ属・ツマグロゼミ属 [セミ科] (半翅目).
    Cryptotympana spp. & *Vagitanus* spp. [Cicadidae] (Hemiptera).

11

12

13

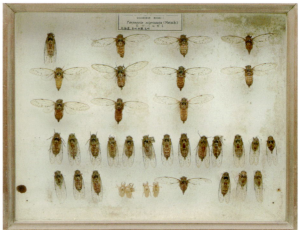
14

034

11. クサゼミの翅脈相調査標本［セミ科］（半翅目）．
    Wing venation pattern of *Mogannia hebes* [Cicadidae] (Hemiptera).

12. カレイゼミ・ヤエヤマクマゼミ・ホソヒグラシ他［セミ科］（半翅目）．
    Macrosemia kareisana, *Cryptotympana yayeyamana*, *Leptosemia sakaii*, etc.
    [Cicadidae] (Hemiptera).

13. ハルゼミ［セミ科］（半翅目）．
    Terpnosia vacua [Cicadidae] (Hemiptera).

14. エゾハルゼミ［セミ科］（半翅目）．
    Terpnosia nigricosta [Cicadidae] (Hemiptera).

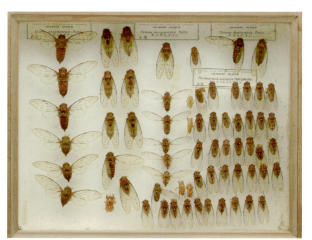
15

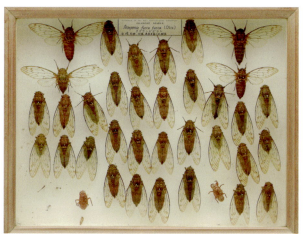
16

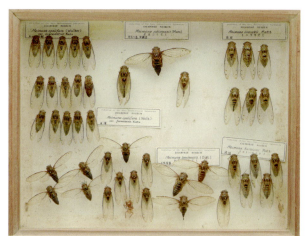
17

18

15. ヒグラシ属・タイワンヒグラシ属 [セミ科] (半翅目).
    Tanna spp. & *Pomponia* spp. [Cicadidae] (Hemiptera).

16. タイワンヒグラシ [セミ科] (半翅目).
    Pomponia yayeyamana [Cicadidae] (Hemiptera).

17. オオシマゼミ・オガサワラゼミ・クロイワツクツク他 [セミ科] (半翅目).
    Meimuna oshimensis, M. boninensis, M. kuroiwae, etc. [Cicadidae] (Hemiptera).

18. タカネゼミ (カレイゼミ)・タカサゴゼミ・ツクツクボウシ [セミ科] (半翅目).
    Cosmopsaltria montana (= *Macrosemia kareisana*), *Dlatylomia bivocalis* & *Meimuna opalifera* [Cicadidae] (Hemiptera).

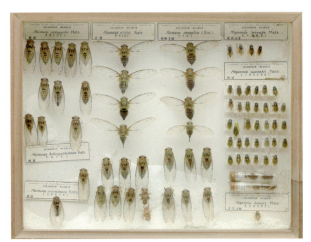
19

20

21

22

19. ウライゼミ（イワサキゼミ）・コマゼミ・イワサキクサゼミ他［セミ科］（半翅目）.
 Meimuna uraina (= *Meimuna iwasakii*), *Meimuna mongolica*, *Mogannia minuta*, etc. [Cicadidae] (Hemiptera).

20. ミンミンゼミ［セミ科］（半翅目）.
 Hyalessa maculaticollis [Cicadidae] (Hemiptera).

21. アブラゼミ・リュウキュウアブラゼミ［セミ科］（半翅目）.
 Graptopsaltria nigrofuscata & *G. bimaculata* [Cicadidae] (Hemiptera).

22. タイワンアブラゼミ［セミ科］（半翅目）.
 Formotosena seebohmi [Cicadidae] (Hemiptera).

23

24

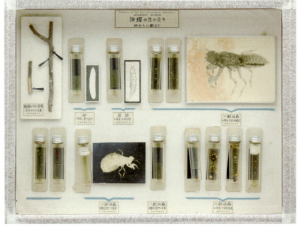

25

26

23. チッチゼミ［セミ科］（半翅目）.
    Kosemia radiator [Cicadidae] (Hemiptera).

24. アシアカハグロゼミ・エゾチッチゼミ他［セミ科］（半翅目）.
    Scieroptera formosana, *Kosemia yezoensis*, etc. [Cicadidae] (Hemiptera).

25. セミの産卵.
    Oviposition of cicadas.

26. アブラゼミの卵から2齢.
    Eggs to 2nd instar larvae of *Graptopsaltria nigrofuscata*.

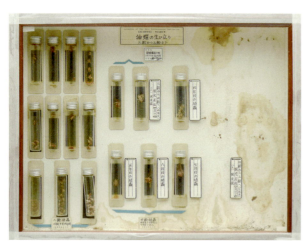
27

28

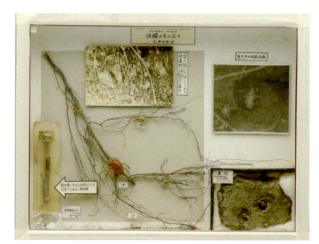
29

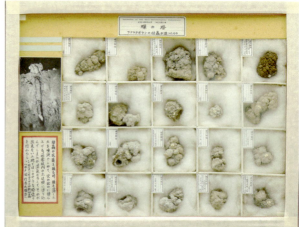
30

27. アブラゼミの3齢から5齢.
    Third to 5th instar larvae of *Graptopsaltria nigrofuscata*.

28. アブラゼミの羽化.
    Eclosion of *Graptopsaltria nigrofuscata*.

29. アブラゼミ幼虫の生活.
    Underground-life of larvae in *Graptopsaltria nigrofuscata*.

30. ツクツクボウシの幼虫が造った蝉塔.
    Mud towers constructed by larvae of *Meimuna opalifera*.

31

33

32

34

31. セミヤドリガ.
    Epipomponia nawai.

32. セミヤドリガの幼虫.
    Larvae of *Epipomponia nawai.*

33. モズの早贄.
    Cicadas and a bumblebee impaled on twigs by a shrike.

34. セミの天敵.
    Natural enemies of cicadas.

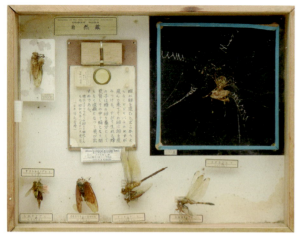
35

36

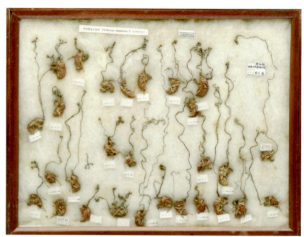
37

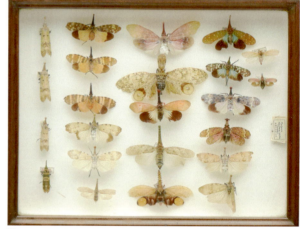
38

35. セミの天敵.
    Natural enemies of cicadas.

36. セミの保護色・警戒色・擬態.
    Cryptic coloration, warning colouration and mimicry of cicadas.

37. アブラゼミタケ[オフィオコルディケプス科](ボタンタケ目).
    Polycephalomyces nipponicus [Ophiocordycipitaceae] (Hypocreales).

38. ビワハゴロモ類[ビワハゴロモ科](半翅目).
    Lanternflies [Fulgoridae] (Hemiptera).

39

40

41

42

39. ビワハゴロモ類［ビワハゴロモ科］（半翅目）.
    Lanternflies [Fulgoridae] (Hemiptera).

40. タケウチトゲアワフキ［トゲアワフキムシ科］（半翅目）.
    Machaerota takeuchii [Machaerotidae] (Hemiptera).

41. カメムシ類（半翅目）.
    Stink bugs (Hemiptera).

42. カメムシ類（半翅目）.
    Stink bugs (Hemiptera).

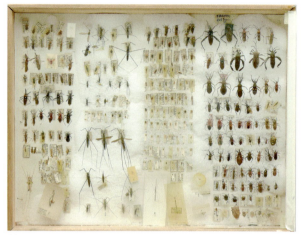

43

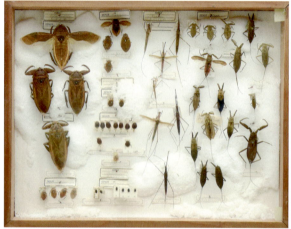

44

45

46

43. カメムシ類・アメンボ類（半翅目）.
    Stink bugs & water striders (Hemiptera).

44. 東京産タガメ・コバンムシ他［タイコウチ下目］（半翅目）.
    Lethocerus deyrollei, Ilyocoris cimicoides, etc. from Tokyo [Hydrocorisae] (Hemiptera).

45. コオニヤンマ・オナガサナエ他［サナエトンボ科］（蜻蛉目）.
    Sieboldius albardae, Melligomphus viridicostus, etc. [Gomphidae] (Odonata).

46. 石神井産ベッコウトンボ・ヨツボシトンボ他［トンボ科］（蜻蛉目）.
    Libellula angelina, Libellula quadrimaculata, etc. from Shakujii, Tokyo [Libellulidae] (Odonata).

47

49

48

50

47. 東京産アカハネバッタ・カワラバッタ他［バッタ科］（直翅目）．
   Celes akitanus, Eusphingonotus japonicus, etc. from Tokyo [Acrididae] (Orthoptera).

48. 横浜産ヒメシロチョウ・ミヤマシロチョウ・マダラシロチョウ他［シロチョウ科］（鱗翅目）．
   Leptidea amurensis from Yokohama, *Aporia hippia*, *Prioneris thestylis*, etc. [Pieridae] (Lepidoptera).

49. 東京産オオウラギンヒョウモン・ウラギンヒョウモン他［タテハチョウ科］（鱗翅目）．
   Fabriciana nerippe, *Fabriciana adippe*, etc. from Tokyo [Nymphalidae] (Lepidoptera).

50. ドクガ［ドクガ科］（鱗翅目）．
   Artaxa subflava [Lymantriidae] (Lepidoptera).

51

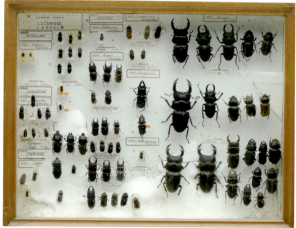
53

52

54

51. ガ類の幼虫.
    Larvae of moths.

52. オオアワダチソウを訪れたハチ類(石神井産).
    Bees visiting *Solidago gigantea* in Shakujii, Tokyo.

53. カブトムシの変態.
    Metamorphosis of the Japanese rhinocerous beetle, *Trypoxylus dichotomus*.

54. クワガタ類[クワガタムシ科] (鞘翅目).
    Stag beetles [Lucanidae] (Coleoptera).

55

56

57

55. クモ類（蜘蛛目）.
    Spiders (Araneae).

56. クモ類（蜘蛛目）.
    Spiders (Araneae).

57. 昆虫玩具.
    Antique insect toys.

1-57. 箱寸法 (Box dimantions) 302×394 mm.

045

山階芳麿
──山階鳥類研究所の創立者

Yoshimaro Yamashina
— Founder of the Yamashina Institute for Ornithology

山階芳麿（1900〜1989）

山階芳麿博士（1900〜1989年）は著名な鳥類の研究者である。山階宮菊麿王の第二王子として誕生、陸軍士官学校を経て陸軍に入ったが、幼少の頃より興味を持ち続けた鳥類の研究に専念するため退役。東京帝国大学理学部動物学科選科修了後、1932年に山階家鳥類標本館を設立、1942年に後身の財団法人山階鳥類研究所の初代所長となり、鳥類の研究と保護にその後の生涯を捧げた。1942年北海道帝国大学理学博士。日本鳥学会会頭、日本鳥類保護連盟会長、国際鳥類保護会議副会長などを歴任。1977年ジャン・デラクール賞、1978年オランダ王室第1級ゴールデンアーク勲章を受賞。

山階家鳥類標本館の設立前後、博士はアジアの動物相調査に精力的に取り組み、今回展示する昆虫標本コレクションも主にこのときに作り上げられた。調査はいつも「三段構え」で行われたという。まず最初に、標本採集者として名高い折居彪二郎氏が調査地に派遣される。彼は現地に1、2年住み込んで採集を行い、作成した標本を博士に送る。新たに採れる動物がいなくなってきたところで、専門のトレーニングを受けた山田信夫氏（ときには日和三徳氏、鳥居元氏）が現地に年単位で滞在、生態の調査を行った。最後に博士自らが現地を訪れ、基礎調査の結果を一つ一つ確認して回ったという。

このときの調査やその後に得た標本をもとに博士は2属3種40亜種の鳥類を命名した。博士が集めた哺乳類や昆虫類などの標本はそれぞれの分類群の専門家によって精査され、新種・新亜種の報告の際、証拠標本として用いられた。特に昆虫標本は主に鱗翅目（チョウ・ガ類）で構成され、一部セミ類なども見られる。中には北海道大学教授を務めていた松村松年博士により1939年に記載された満州産チョウ類3新種25新型（新亜種）のタイプ標本も含まれている。また、このコレクションには、山階博士の御父君・山階宮菊麿王殿下によって採集されたチョウ類標本も見受けられる。

博士の遺した標本コレクションは、現在の公益財団法人山階鳥類研究所に受け継がれている。鳥類のコレクションは、国内最大の標本点数を誇っている。一方、昭和初期のアジアの資料として高い価値を備える昆虫標本コレクション約140箱は、2017年冬に東京大学総合研究博物館へ寄贈された。

（山崎剛史）

Dr. Yoshimaro Yamashina (1900 – 1989) was a prominent ornithologist. Although he was born into royalty as the second son of Imperial Prince Yamashinanomiya Kikumaro, enjoying a prestigious position in the Imperial Japanese Army, he left both his title and position to study ornithology, his passion since childhood. After graduating from the Department of Zoology at Tokyo Imperial University, he founded Yamashina's Private Laboratory Museum in 1932. He later went on to found the Yamashina Institute for Ornithology in 1942, at which he devoted his life to the research and conservation of birds. Also in 1942, he received a doctorate of science from Hokkaido Imperial University. During his lifetime, he served as President of the Ornithological Society of Japan, President of the Japanese Society for Preservation of Birds, and Vice President of the International Council for Bird

Preservation. He was awarded the Jean Delacour Prize in 1977, and in 1978 received the Royal Order of the Golden Ark, the Netherlands' highest conservation honor.

Around the time of the foundation of the Yamashina's Private Laboratory Museum, Dr. Yamashina worked tirelessly on a comprehensive investigation of Asian fauna, during which he made insect specimens as well as bird specimens. The field research is said to have generally been conducted in three steps. First, Mr. Hyojiro Orii, a famous wildlife collector, was dispatched to a research site where he would stay for a lengthy period, sending specimens back to Dr. Yamashina for further analysis. Subsequently, Mr. Nobuo Yamada (or occasionally Mr. Minori Hiwa or Mr. Hajime Torii), a special researcher, would replace Mr. Orii to research the local wildlife ecology for a year. Finally, Dr. Yamashina would visit the research site and verify in person the results of the basic survey.

Based on specimens obtained from the research, he originally described two new genera, three new species and 40 new subspecies of birds. His animal and insect specimens were carefully examined by experts in their respective taxonomic fields, and were later used as voucher specimens in publishing new species and subspecies. The insect specimens are composed mainly of Lepidoptera (butterflies and moths) and partly cicadas. Of the butterfly specimens, the holotype of three new species and 25 new forms (currently subspecies) from Manchuria, described by Prof. Shonen Matsumura (Hokkaido University) in 1939, are included. In addition, his collection also contains several butterfly specimens collected by his father, Prince Yamashinanomiya Kikumaro.

Dr. Yamashina's collection was inherited by the Yamashina Institute for Ornithology, where the bird and animal collections currently reside. The bird collection is remarkable for being the largest in Japan. His insect collection was kindly deposited to the University Museum, the University of Tokyo in the winter of 2017. The insect collection, which comprises roughly 140 cabinets, is highly valuable due to the large number of Asian specimens collected in the early Showa era.

(Takeshi Yamasaki)

1

2

1. アカエゾゼミ・ミンミンゼミ他［セミ科］（半翅目）.
   Lyristes flammatus, *Hyalessa maculaticollis*, etc. [Cicadidae] (Hemiptera).

2. アブラゼミ・クマゼミ他［セミ科］（半翅目）.
   Grapopsaltria nigrofuscata, *Cryptotympana facilialis*, etc. [Cicadidae] (Hemiptera).

5

6

3

4

048

3. 東京産ギフチョウ［アゲハチョウ科］・奈良産ルーミスシジミ［シジミチョウ科］・父島産オガサワラセセリ［セセリチョウ科］他（鱗翅目）.
Luehdorfia japonica [Papilionidae] from Tokyo, *Arhopala ganesa* [Lycaenidae] from Nara, *Parnara ogasawarensis* [Hesperiidae] from Chichi-jima Island, etc. (Lepidoptera).

4. キアゲハ［アゲハチョウ科］・ウチスズメ［スズメガ科］他（鱗翅目）.
Papilio machaon [Papilionidae], *Smerinthus planus* [Sphingidae], etc. (Lepidoptera).

5. アオスジアゲハ・モンキアゲハ他［アゲハチョウ科］他（鱗翅目）.
Graphium sarpedpn & *Papilio helenus* [Papilionidae], etc. (Lepidoptera).

6. オナガアゲハ［アゲハチョウ科］・コムラサキ［タテハチョウ科］他（鱗翅目）.
Papilio macilientus [Papilionidae], *Apatura metis* [Nynphalidae], etc. (Lepidoptera).

7. カラスアゲハ［アゲハチョウ科］・オオゴマダラ［タテハチョウ科］他（鱗翅目）.
   Papilio dehaanii [Papilionidae], *Kalima inachus* [Nymphalidae], etc. (Lepidoptera).

8. ミヤマカラスアゲハ［アゲハチョウ科］・ヒメシロチョウ［シロチョウ科］他（鱗翅目）.
   Papilio maackii [Papilionidae], *Leptidea amurensis* [Pieridae], etc. (Lepidoptera).

9. ベニモンシロチョウ・タイワンシロチョウ他［シロチョウ科］（鱗翅目）.
   Delias hyparete, *Appias lyncida*, etc. [Pieridae] (Lepidoptera).

10. ツマベニチョウ・タイワンメスシロキチョウ他［シロチョウ科］（鱗翅目）.
    Hebomoia graucippe, *Ixias insignis*, etc. [Pieridae] (Lepidoptera).

11

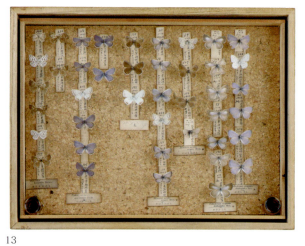
13

12

14

050

11. エゾシロチョウ［シロチョウ科］・キバネセセリ［セセリチョウ科］他（鱗翅目）.
   Aporia cratagei [Pieridae], *Bibasis aqulina* [Hesperiidae], etc. (Lepidoptera).

12. モンキチョウ・キタキチョウ［シロチョウ科］（鱗翅目）.
   Colias erate & *Eurema mandarina* [Pieridae] (Lepidoptera).

13. ゴイシシジミ・ツバメシジミ他［シジミチョウ科］（鱗翅目）.
   Taraka hamada, *Everes argiades*, etc. [Lycaenidae] (Lepidoptera).

14. アカシジミ［シジミチョウ科］オオヒカゲ［タテハチョウ科］他（鱗翅目）.
   Japonica lutea [Lycaenidae], *Ninguta schrenckii* [Nymphalidae], etc. (Lepidoptera).

15

17

16

18

15. キベリタテハ・オオイチモンジ［タテハチョウ科］他（鱗翅目）.
    Nymphalis antiopa, Limenitis populi [Nymphalidae], etc. (Lepidoptera).

16. アオタテハモドキ・タテハモドキ他［タテハチョウ科］（鱗翅目）.
    Junonia orithya, Junonia almana, etc. [Nymphalidae] (Lepidoptera).

17. イワサキタテハモドキ・ジャノメタテハモドキ［タテハチョウ科］他（鱗翅目）.
    Junonia hedonia, J. lemonias [Nymphalidae], etc. (Lepidoptera).

18. ヤエヤマムラサキ・ヤホシムラサキ［タテハチョウ科］他（鱗翅目）.
    Hypolimnas anoamla & *Hypolimnas octocula* [Nymphalidae], etc. (Lepidoptera).

19

20

21

22

052

19. リュウキュウムラサキ［タテハチョウ科］（鱗翅目）.
    Hypolimnas bolina [Nymphalidae] (Lepidoptera).

20. リュウキュウムラサキ・マルバネルリマダラ他［タテハチョウ科］（鱗翅目）.
    Hypolimnas bolina, Euploea eunice, etc. [Nymphalidae] (Lepidoptera).

21. タカサゴイチモンジ・コノハチョウ他［タテハチョウ科］（鱗翅目）.
    Euthalia formosana, Kallima inachus, etc. [Nymphalidae] (Lepidoptera).

22. ウラギンヒョウモン・ヒョウモンチョウ他［タテハチョウ科］（鱗翅目）.
    Argynnis adippe, Brenthis daphne, etc. [Nymphalidae] (Lepidoptera).

23

24

25

26

23. ミドリヒョウモン［タテハチョウ科］（鱗翅目）.
　　Arginnis paphia [Nymphalidae] (Lepidoptera).

24. ギンボシヒョウモン・ウラギンスジヒョウモン他［タテハチョウ科］（鱗翅目）.
　　Arginnis aglaja, Argironome laodice, etc. [Nymphalidae] (Lepidoptera).

25. メスグロヒョウモン・オオウラギンスジヒョウモン［タテハチョウ科］（鱗翅目）.
　　*Damora sagan*a & *Argyronome ruslana* [Nymphalidae] (Lepidoptera).

26. メスグロヒョウモン・キゴマダラ他［タテハチョウ科］（鱗翅目）.
　　Damora sagana, Sephisa chandra, etc. [Nymphalidae] (Lepidoptera).

27

28

29

30

27. オオミスジ［タテハチョウ科］・コキマダラセセリ［セセリチョウ科］他（鱗翅目）．
    Neptis alwina [Nymphalidae], *Ochlodes venatus* [Hesperiidae], etc. (Lepidoptera).

28. 箱根産アサマイチモンジ・イチモンジチョウ他［タテハチョウ科］（鱗翅目）．
    Limenitis glorifica & *Limenitis camilla* from Hakone, etc. [Nymphalidae] (Lepidoptera).

29. オオカバマダラ［タテハチョウ科］（鱗翅目）．
    Danaus plexippus [Nymphalidae] (Lepidoptera).

30. カバマダラ・ツマムラサキマダラ他［タテハチョウ科］（鱗翅目）．
    Danaus chrysippus, *Euploea mulciber*, etc. [Nymphalidae] (Lepidoptera).

31

32

33

34

31. ヒカゲチョウ[タテハチョウ科]・アゲハモドキ[アゲハモドキガ科]他(鱗翅目).
    Lethe sicelis [Nymphalidae], *Epicopeia hainesii* [Epicopeiidae], etc. (Lepidoptera).

32. サトキマダラヒカゲ・ヒメキマダラヒカゲ他[タテハチョウ科](鱗翅目).
    Neope gaschkevitschii, Zophoessa callipteris, etc. [Nymplalidae] (Lepidoptera).

33. 箱根産ウラナミジャノメ・ヒメウラナミジャノメ・ジャノメチョウ他[タテハチョウ科](鱗翅目).
    Yptima multistriata, Y. argus & *Minois dryas* from Hakone [Nymphalidae], etc. (Lepidoptera).

34. ウスイロコノマチョウ・ルリモンジャノメ他[タテハチョウ科](鱗翅目).
    Melanitis leda, Elymnias hypermnestra, etc. [Nymphalidae] (Lepidoptera).

35

36

37

38

35. ワモンチョウ［タテハチョウ科］・ウラギンシジミ［シジミチョウ科］他（鱗翅目）．
    Stichophthalma howqua [Nymphalidae], *Curetis acuta* [Lycaenidae], etc. (Lepidoptera).

36. ダイミョウセセリ・コチャバネセセリ他［セセリチョウ科］（鱗翅目）．
    Daimio tethys, Thoressa varia, etc. [Hesperiidae] (Lepidoptera).

37. 箱根産チャマダラセセリ・ホソバセセリ・コキマダラセセリ他［セセリチョウ科］（鱗翅目）．
    Pyrgus maculatus, Isoteinon lamprospilus & *Ochlodes venatus* from Hakone [Hespariidae], etc. (Lepidoptera).

38. ベニスズメ・ウンモンスズメ他［スズメガ科］（鱗翅目）．
    Deilephila elpenor, Callanbulyx tatarinovii, etc. [Sphingidae] (Lepidoptera).

39

40

41

42

43

39. アカスジシロコケガ・ゴマダラベニコケガ他［ヒトリガ科］（鱗翅目）.
    Cyana hamata, Barsine purchra, etc. [Erebidae] (Lepidoptera).

40. ホシベニシタヒトリ［ヒトリガ科］・モモスズメ［スズメガ科］他（鱗翅目）.
    Ryparia amurensis [Erebidae], *Marumba gaschkewitschii* [Sphingidae], etc. (Lepidoptera).

41. チズモンアオシャク・ホシスジトガリナミシャク他［シャクガ科］（鱗翅目）.
    Agathia visenda, Carige cruciplaga, etc. [Geometridae] (Lepidoptera).

42. ヨツメエダシャク・クロクモエダシャク他［シャクガ科］（鱗翅目）.
    Ophthalmitis albosignaria, Apocleora rimosa, etc. [Geometridae] (Lepidoptera).

43. ムラサキシタバ・ムクゲコノハ他［ヤガ科］（鱗翅目）.
    Catocala fraxini, Thyas juno, etc. [Noctuidae] (Lepidoptera).

1-43. 箱寸法 (Box dimantions) 285×376 mm.

第 2 章

昭和中期［戦後］〜平成

Chapter 2
Mid-Showa to Heisei eras

五十嵐　邁
―― 蝶類幼生期研究の大家

Suguru Igarashi
— An authority on the early stages of butterflies

五十嵐　邁（1924〜2008）

五十嵐邁博士（1924〜2008年）はチョウ類の幼生期の生活史と形態を主に研究した昆虫学者として知られる。長崎県佐世保で出生後、1947年に東京大学工学部を卒業し、大成建設（株）取締役（1979〜1985年）、信越半導体（株）取締役社長（1990〜1996年）などを務めた実業家でもあり、作家としても偉業を残す。最も有名な業績は、幼虫期の生態や形態が不明だった珍種テングアゲハの調査団を結成して、インドのダージリンで食樹がキャンベリーモクレンであることを発見し、幼生期を解明したことであろう。彼の研究は「世界のアゲハチョウ．講談社（1979）」としてまとめられ、1983年に京都大学で理学博士の学位も取得した。日本蝶類学会の初代会長（1992〜1995年）や名誉会長（1999〜2008年）も歴任している。

五十嵐コレクションは、タイプ標本を含む東南アジア産チョウ類を主とする約10万点のチョウ類標本、1000点を優に越える学術図書、5000点以上のチョウ類幼生期の写真や描図などである。この中には現在取引できない種、世界的にも極めて貴重なチョウ類標本やタイプ標本などが見られる。これらの代表的な種として、アレクサンドラトリバネアゲハ、ルソンカラスアゲハ、オナシカラスアゲハ、ランプサックスアゲハなどが挙げられる。自身の大図鑑等で使用された多数の原図や昆虫史に名が残る故・磐瀬太郎氏の書籍、トンボやゴキブリの分類・生態学的研究の第一人者だった国立予防衛生研究所の朝比奈正二郎博士の満州産チョウ類標本なども含まれる。東京大学へのコレクション寄贈は、同博士の奥方・昌子氏の意向で2010年6月に実現した。この寄贈の功績が評価され、同年秋に昌子氏には本学から功績者顕彰制度「稷門賞」も受賞されている。

（矢後勝也）

Dr. Suguru Igarashi (1924–2008), born in Sasebo, Nagasaki, was a well-known amateur entomologist who mainly studied the life histories and early stage morphology of butterflies. He studied at the Faculty of Engineering, the University of Tokyo, after which he went on to not only hold many high-ranking business positions, including a board member of the Taisei Corporation (1979–1985) and the president of Shin-Etsu Handotai Co., Ltd. (1990–1996), but also to become an accomplished novelist. His most famous achievement is that he formed a research team that led to the discovery of the immature stages of a rare papilionid, *Teinopalpus imperialis*, in Darjeeling, India. In 1979, his work culminated in a publication entitled "Papilionidae and their Early Stages" (Kodansha), for which he received a doctorate of science from Kyoto University in 1983. He was notably the first president (1992–1995) and president emeritus of the Butterfly Society of Japan (1999–2008).

Most of his collection is related to the life histories of butterflies: about 1,000,000 specimens including the holotype of some butterflies, more than 1,000 books and journals and more than 5,000 paintings and photographs of their early stages. Among this great collection include species on the current embargo list, very rare and valuable species known elsewhere in the world from only limited specimens, and original type specimens of taxa which he described. Their representative species are *Ornithoptera alexandrae*,

Achillides chikae, *Achillides elephenor*, *Menelaides lampsacus* and others. Moreover, the collection contains original illustrations which were published as figures in his butterfly books, books owned by the late Mr. Tarô Iwase (a legendary figure in the history of Japanese lepidopterology), and Manchurian butterfly specimens collected by the late Dr. Shojiro Asahina who studied taxonomy and biology of dragonflies and cockroaches at the National Institute of Health, Japan. The donation of his collection to The University of Tokyo was carried out in June, 2010 according to the wishes of his wife, Mrs. Yoshiko Igarashi. In the autumn of 2010, Mrs. Igarashi was awarded the "Shokumon-prize" from the university in recognition of the contribution.

(Masaya Yago)

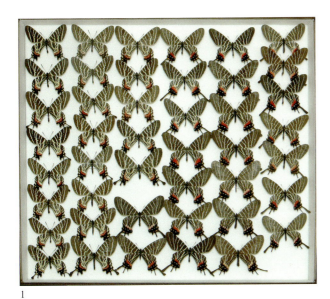

1

2

3

4

1. ウンナンシボリアゲハ・シナシボリアゲハ［アゲハチョウ科］（鱗翅目）.
   Bhutanitis mansfieldi & *B. thaidina* [Papilionidae] (Lepidoptera).

2. シボリアゲハ［アゲハチョウ科］（鱗翅目）.
   Bhutanitis lidderdalii [Papilionidae] (Lepidoptera).

3. ヒマラヤアケボノアゲハ［アゲハチョウ科］（鱗翅目）.
   Atrophaneura aidoneus [Papilionidae] (Lepidoptera).

4. キイロハゲタカアゲハ［アゲハチョウ科］（鱗翅目）.
   Atrophaneura priapus [Papilionidae] (Lepidoptera).

5

6

7

8

062

5. ミランダキシタアゲハ [アゲハチョウ科] (鱗翅目).
   Troides miranda [Papilionidae] (Lepidoptera).

6. アカメガネトリバネアゲハ [アゲハチョウ科] (鱗翅目).
   Ornithoptera croesus [Papilionidae] (Lepidoptera).

7. アレクサンドラトリバネアゲハ [アゲハチョウ科] (鱗翅目).
   Ornithoptera alexandrae [Papilionidae] (Lepidoptera).

8. テングアゲハ [アゲハチョウ科] (鱗翅目).
   Teinopalpus imperialis [Papilionidae] (Lepidoptera).

9

10

11

12

9. フィディアスタイマイ他[アゲハチョウ科] (鱗翅目).
   Paranticopsis phidias [Papilionidae] (Lepidoptera).

10. ジョルダンアゲハ・マハデヴァオナシモンキアゲハ[アゲハチョウ科] (鱗翅目).
    Menelaides jordani & *Menelaides mahadeva* [Papilionidae] (Lepidoptera).

11. ランプサックスアゲハ[アゲハチョウ科] (鱗翅目).
    Menelaides lampsacus [Papilionidae] (Lepidoptera).

12. ルソンカラスアゲハ・ミンドロカラスアゲハ[アゲハチョウ科] (鱗翅目).
    Achillides chikae & *A. hermeli* [Papilionidae] (Lepidoptera).

13

14

15

16

13. オナシカラスアゲハ・タイワンカラスアゲハ［アゲハチョウ科］（鱗翅目）.
    Achillides elephenor & *A. dialis* [Papilionidae] (Lepidoptera).

14. タカネクジャクアゲハ［アゲハチョウ科］（鱗翅目）.
    Achillides krishna [Papilionidae] (Lepidoptera).

15. カルナルリモンアゲハ［アゲハチョウ科］（鱗翅目）.
    Achillides karna [Papilionidae] (Lepidoptera).

16. スンバアオネアゲハ・ブッダオビクジャクアゲハ［アゲハチョウ科］（鱗翅目）.
    Achillides neumoegeni & *Achillides buddha* [Papilionidae] (Lepidoptera).

17

18

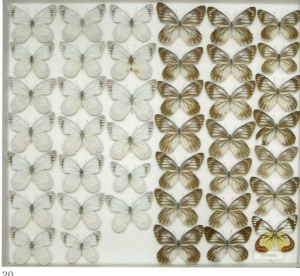

19

20

17. オオルリオビアゲハ［アゲハチョウ科］(鱗翅目).
    Achillides blumei [Papilionidae] (Lepidoptera).

18. オオルリアゲハ［アゲハチョウ科］(鱗翅目).
    Achillides ulysses [Papilionidae] (Lepidoptera).

19. ヒイロツマベニチョウ［シロチョウ科］(鱗翅目).
    Hebomoia leucippe [Pieridae] (Lepidoptera).

20. ローゼンバーグカザリシロチョウ［シロチョウ科］(鱗翅目).
    Delias rosenbergi [Pieridae] (Lepidoptera).

21

22

23

24

21. アサギシロチョウ属［シロチョウ科］（鱗翅目）.
    Pareronia spp. [Pieridae] (Lepidoptera).

22. メスシロキチョウ属［シロチョウ科］（鱗翅目）.
    Ixias spp. [Pieridae] (Lepidoptera).

23. テイオウスソビキシジミ・ヒスイスソビキシジミ・ヤドリギオナガシジミ他
    ［シジミチョウ科］（鱗翅目）.
    Purlisa gigatea, Charana mandrina, Tajuria cippus, etc. [Lycaenidae] (Lepidoptera).

24. イワカワシジミ・ヒイロシジミ他［シジミチョウ科］（鱗翅目）.
    Artipe eryx, Deudorix epijarbas, etc. [Lycaenidae] (Lepidoptera).

25

26

27

28

25. フチグロベニシジミ・メスグロベニシジミ・オオベニシジミ他［シジミチョウ科］（鱗翅目）.
  Heodes virgaureae, Palaeochrysophanus hippothoe, Thersamolycaena dispar, etc. [Lycaenidae] (Lepidoptera).

26. タスキシジミ［シジミチョウ科］（鱗翅目）.
  Danis danis [Lycaenidae] (Lepidoptera).

27. ミイロタテハ属・ルリオビタテハ属［タテハチョウ科］（鱗翅目）.
  Agrias spp. & *Prepona* spp. [Nymphalidae] (Lepidoptera).

28. フタオチョウ［タテハチョウ科］（鱗翅目）.
  Polyura weismanni [Nymphalidae] (Lepidoptera).

29

30

31

32

068

29. ヤイロタテハ[タテハチョウ科] (鱗翅目).
　　Agatasa calydonia [Nymphalidae] (Lepidoptera).

30. ルリオビヤイロタテハ[タテハチョウ科] (鱗翅目).
　　Prothoe franck [Nymphalidae] (Lepidoptera).

31. モンキコムラサキ・マドタテハ・イチモンジマドタテハ他[タテハチョウ科]
　　(鱗翅目).
　　Euapatura mirza, Dilipa fenestra, Lelecella limenitoides, etc. [Nymphalidae]
　　(Lepidoptera).

32. オオムラサキ[タテハチョウ科] (鱗翅目).
　　Sasakia charonda [Nymphalidae] (Lepidoptera).

33.

34.

35.

36.

33. アカネゴマダラ・エグリゴマダラ［タテハチョウ科］（鱗翅目）.
 Euripus consimilis & *Euripus nyctelius* [Nymphalidae] (Lepidoptera).

34. ベニボシイナズマ・アドニアベニボシイナズマ・イナズマチョウ他［タテハチョウ科］（鱗翅目）.
 Euthalia lubentina, Euthalia adonia, Euthalia irrubescens, etc. [Nymphalidae] (Lepidoptera).

35. ハレギチョウ属［タテハチョウ科］（鱗翅目）.
 Cethosia spp. [Nymphalidae] (Lepidoptera).

36. コノハチョウ［タテハチョウ科］（鱗翅目）.
 Kallima inachus [Nymphalidae] (Lepidoptera).

37

38

39

40

37. マネシジャノメ属［タテハチョウ科］（鱗翅目）．
    Elymnias spp. [Nymphalidae] (Lepidoptera).

38. シロスジマダラ・オオシロスジマダラ［タテハチョウ科］（鱗翅目）．
    Penthema formosanum & *P. darlisa* [Nymphalidae] (Lepidoptera).

39. ルリヒカゲ属［タテハチョウ科］（鱗翅目）．
    Ptychandra spp. [Nymphalidae] (Lepidoptera).

40. コジャノメ属［タテハチョウ科］（鱗翅目）．
    Mycalesis spp. [Nymphalidae] (Lepidoptera).

41

42

43

44

071

41. シロオビワモンチョウ属 [タテハチョウ科] (鱗翅目).
    Thauris spp. [Nymphalidae] (Lepidoptera).

42. ワモンチョウ属 [タテハチョウ科] (鱗翅目).
    Stichophthalma spp. [Nymphalidae] (Lepidoptera).

43. タイヨウモルフォ・シロモルフォ・フクロウチョウ他 [タテハチョウ科] (鱗翅目).
    Morpho hecuba, Morpho polyphemus, Caligo teucer, etc. [Nymphalidae] (Lepidoptera).

44. メネラウスモルフォ・ディディウスモルフォ [タテハチョウ科] (鱗翅目).
    Morpho menelaus & *M. didius* [Nymphalidae] (Lepidoptera).

45

46

47

48

072

45. ウスグロゴマダラ・タンブシシオオゴマダラ・オオゴマダラ他［タテハチョウ科］（鱗翅目）．
    Idea blanchardii, I. tambusisiana, I. leuconoe, etc. [Nymphalidae] (Lepidoptera).

46. キイロアサギマダラ・スケンクアサギマダラ・ビトリナヒメゴマダラ他［タテハチョウ科］（鱗翅目）．
    Parantica cleona, P. schenkii, P. vitrina, etc. [Nymphalidae] (Lepidoptera).

47. オオムラサキマダラ［タテハチョウ科］（鱗翅目）．
    Euploea phaenareta & *E. callithoe* [Nymphalidae] (Lepidoptera).

48. シロモンルリマダラ・チビムラサキマダラ・シロモンチビマダラ他［タテハチョウ科］（鱗翅目）．
    Euploea radamantus, E. eleusina, Tellervo zoilus, etc. [Nymphalidae] (Lepidoptera).

49

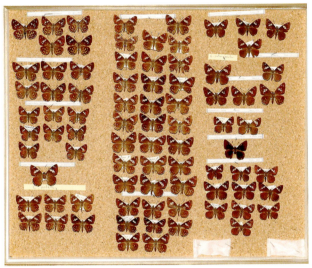

50

51

073

49. アオバセセリ属 [セセリチョウ科] (鱗翅目).
　　Choaspes spp. [Hesperiidae] (Lepidoptera).

50. キコモンセセリ属 [セセリチョウ科] (鱗翅目).
　　Celaenorrhinus spp. [Hesperiidae] (Lepidoptera).

51. アカメセセリ属・ヤシセセリ属・ウラジロセセリ属他. [セセリチョウ科]
　　(鱗翅目).
　　Matapa spp., *Hidari* spp., *Unkana* spp., etc. [Hesperiidae] (Lepidoptera).

1-51. 箱寸法 (Box dimantions) 418×507 mm.

江田　茂
——国内有数の大収集家

Shigeru Eda
— One of the greatest insect collectors in Japan

江田　茂（1930～2008）

江田　茂氏（1930～2008年）は、日本有数の昆虫標本の大収集家として知られている。東京で生まれた江田氏は、幼少期を過ごした兵庫県芦屋市にて昆虫採集の魅力に目覚める。中学生の頃より、堪能な英語力で海外の昆虫標本商や研究者、収集家と文通、交換をして、珍奇で美麗なチョウや甲虫の標本コレクションを始めた。灘高校卒業後、東京大学法学部に入学する。その後は1954年に労働省入省、広島県民生労働部職業安定課長などを歴任後、大臣官房統計情報部長を最後に退官した。これらの経歴、歳月を重ねる度に、昆虫標本のコレクション熱は、ますます盛んになった。

多くの海外の地にも赴任し、世界各地の博物館、研究者、収集家、昆虫標本商との交流を深め、所蔵標本は膨大な点数に及んだ。4年間のアメリカ合衆国日本大使館勤務（ワシントンD.C.）時代には、世界一の収蔵を誇るスミソニアン博物館との標本交換により、さらにコレクションが充実する。1983年の労働省退官後も精力的に活動を続け、世界各地の昆虫標本を多岐に渡り収集した。

そのコレクションの範囲は、主にチョウ、ガからなる鱗翅目とカブトムシ、クワガタ、コガネムシ、タマムシ、カミキリなどを含む鞘翅目だが、それだけには留まらず、昆虫のあらゆる分類群に及んでいる。彼のコレクションは、主に兵庫県立人と自然の博物館に収められているが、晩年まで所蔵していたコレクションの一部は、逝去後の2009年秋、東京大学総合研究博物館に収蔵されている。

（矢後勝也）

Mr. Shigeru Eda (1930–2008), born in Tokyo, is remembered as one of the greatest insect collectors in Japan. As a child in Ashiya City, Hyogo, he was fascinated with collecting insects. From his junior high school days, his command of English allowed him to exchange insect specimens with foreign insect dealers, as well as with researchers and collectors, and collected many specimens of rare and beautiful butterflies and beetles. After graduating from Nada High School, he was admitted to the Faculty of Law, the University of Tokyo. In 1954, he joined the Ministry of Labour, where he went on to hold a number of senior positions, such as a section chief in the Consumer and Labour Division of Hiroshima Pref. and finally a division manager for the Minister's Secretariat. Over the years he became more enthusiastic with collecting insects.

While at the ministry, he was transferred overseas many times and exchanged actively with foreign museums, researchers, collectors and dealers. During a four-year period at the Japanese embassy in Washington, D.C., he exchanged many specimens with the Smithsonian National Museum of Natural History, which has the largest insect collection in the world. After his retirement in 1983, he continued to collect with considerable energy, amassing numerous insect specimens from all over the world.

His collection is comprised of many insect orders, especially Lepidoptera such as butterflies and moths and Coleoptera such as unicorn beetles, stag beetles, gold beetles, jewel beetles and longhorn beetles. Although much of the material remains housed in the Museum of Nature and Human Activities, Hyogo, many specimens were deposited in the University Museum, the University of Tokyo, in the autumn of 2009.

(Masaya Yago)

1

2

3

4

075

1. ハナカマキリ［ハナカマキリ科］（蟷螂目）・アジアオオキバヘビトンボ［ヘビトンボ科］（脈翅目）他.
   Hymenopus coronatus [Hymenopodidae] (Mantodea), *Acanthacorydalis orientalis* [Corydalidae] (Neuroptera), etc.

2. センストビナナフシ他［ナナフシ科］（竹節虫目）.
   Tagesoidea nigrofasciata, etc. [Phasmatidae] (Phasmida).

3. カービーオオナナフシ［ナナフシ科］（竹節虫目）.
   Phobaeticus kirbyi [phasmatidae] (Phasmida).

4. サカダチコノハナナフシ［サカダチコノハナナフシ科］・オオコノハムシ［コノハムシ科］他（竹節虫目）.
   Heteropteryx dilatata [Heteropterygidae], *Phyllium giganteum* [Phylliidae], etc. (Phasmida).

5

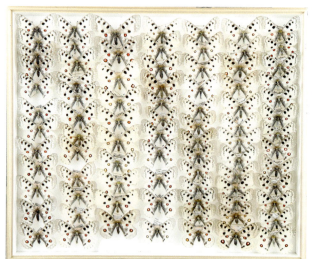

6

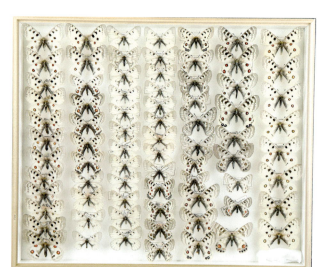

7

8

076

5. モエギクマゼミ・テイオウゼミ［セミ科］他（半翅目）．
   Salvazana mirabilis, Pomponia imperatoria, [Cicadidae], etc. (Hemiptera).

6. アポロウスバシロチョウ［アゲハチョウ科］（鱗翅目）．
   Parnassius apollo [Papilionidae] (Lepidoptera).

7. ケファルスウスバシロチョウ・デルフィウスウスバシロチョウ他
   ［アゲハチョウ科］（鱗翅目）．
   Parnassius cephalus, P. delphius, etc. [Papilionidae] (Lepidoptera).

8. チェケニィウスバシロチョウ・ヒマラヤヒメウスバシロチョウ他
   ［アゲハチョウ科］（鱗翅目）．
   Parnassius szechenyii, P. hardwickii, etc. [Papilionidae] (Lepidoptera).

9

11

10

12

9. フォエブスウスバシロチョウ・ジャケモンウスバシロチョウ他
   [アゲハチョウ科]（鱗翅目）.
   Parnassius phoebus, P. jacquemontii, etc. [Papilionidae] (Lepidoptera).

10. アカボシウスバシロチョウ・アウトクラトールウスバシロチョウ他
    [アゲハチョウ科]（鱗翅目）.
    Parnassius bremeri, P. autocrator, etc. [Papilionidae] (Lepidoptera).

11. ウスバキチョウ・クロホシウスバシロチョウ他［アゲハチョウ科］（鱗翅目）.
    Parnassius eversmanni, P. mnemosyne, etc. [Papilionidae] (Lepidoptera).

12. オオアカボシウスバシロチョウ・ミカドウスバシロチョウ他［アゲハチョウ科］
    （鱗翅目）.
    Parnassius nomion, P. imperator, etc. [Papilionidae] (Lepidoptera).

13

14

15

16

13. キアゲハ・ブレビカウダキアゲハ他［アゲハチョウ科］（鱗翅目）.
    Papilio machaon, P. brevicauda, etc. [Papilionidae] (Lepidoptera).

14. アスカニウスマエモンジャコウアゲハ・キューバマエモンジャコウアゲハ他［アゲハチョウ科］（鱗翅目）.
    Parides ascanius, P. gundlachianus, etc. [Papilionidae] (Lepidoptera).

15. ワルスウェッチアゲハ・カシクスアゲハ・アスコリウスマルバネアゲハ他［アゲハチョウ科］（鱗翅目）.
    Pterourus warscewiczi, P. cacicus, P. ascolius, etc. [Papilionidae] (Lepidoptera).

16. オライアカザリシロチョウ・アポカザリシロチョウ・エリプシスカザリシロチョウ他［シロチョウ科］（鱗翅目）.
    Delias oraia, D. apoensis, D. ellipsis, etc. [Pieridae] (Lepidoptera).

17

18

19

20

17. アレキサンドラモンキチョウ・ミヤマモンキチョウ他［シロチョウ科］（鱗翅目）．
    Colias alexandra, C. palaeno, etc. [Pieridae] (Lepidoptera).

18. ヒイロタテハ・シンジュタテハ・ウサンバラムラサキ他［タテハチョウ科］（鱗翅目）．
    Cymothoe sangaris, Salamis parhassus, Hypolimnas usambara, etc. [Nymphalidae] (Lepidoptera).

19. ルリボカシタテハ・ベニタテハモドキ・キオビオナガアカタテハ他［タテハチョウ科］（鱗翅目）．
    Euphaedra preussi, Precis octavia, Antanartia schaeneia, etc. [Nymphalidae] (Lepidoptera).

20. シロフタオ・モンティエールフタオ・マダガスカルマルバネタテハ他［タテハチョウ科］（鱗翅目）．
    Charaxes lydiae, C. montieri, Euxanthe madagascariensis, etc. [Nymphalidae] (Lepidoptera).

21

22

23

24

21. ウラギンフタオ・フルニエフタオ他［タテハチョウ科］(鱗翅目).
    Charaxes nobilis, C. fournierae, etc. [Nymphalidae] (Lepidoptera).

22. マダガスカルオナガフタオ・ウラギンオオフタオ他［タテハチョウ科］(鱗翅目).
    Charaxes antamboulou, C. andranodorus, etc. [Nymphalidae] (Lepidoptera).

23. プラエネステナルリオビタテハ・デイフィレルリオビタテハ・デモフォンルリオビタテハ他［タテハチョウ科］(鱗翅目).
    Prepona praeneste, P. deiphile, Archaeoprepona demophon, etc. [Nymphalidae] (Lepidoptera).

24. オンファレルリオビタテハ・デモフーンルリオビタテハ他［タテハチョウ科］(鱗翅目).
    Prepona omphale, Archaeoprepona demophoon, etc. [Nymphalidae] (Lepidoptera).

25

27

26

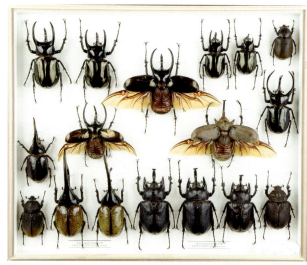

28

25. カラフトタカネヒカゲ・モリシロジャノメ・アリスタエウスタカネジャノメ他 [タテハチョウ科] (鱗翅目).
    Oeneis jutta, Melanargia epimede, Hipparchia aristaeus, etc. [Nymphalidae] (Lepidoptera).

26. エガーモルフォ雌雄型 [タテハチョウ科] (鱗翅目).
    Gynandromorphs of *Morpho aega* [Nymphalidae] (Lepidoptera).

27. ギラファノコギリクワガタ [クワガタムシ科] (鞘翅目)・タイワンタガメ [コオイムシ科] (半翅目)・ベルシルブラオオトビナナフシ [ナナフシ科] (竹節虫目) 他.
    Prosopocoilus giraffa [Lucanidae] (Coleoptera), *Lethocerus indicus* [Belostomatidae] (Hemiptera), *Eurycnema versirubra* [Phasmatinae] (Phasmida), etc.

28. ヘラクレスオオカブト・アクタエオンゾウカブト他 [コガネムシ科] (鞘翅目).
    Dynastes hercules, Megasoma actaeon, etc. [Scarabaeidae] (Coleoptera).

29

30

31

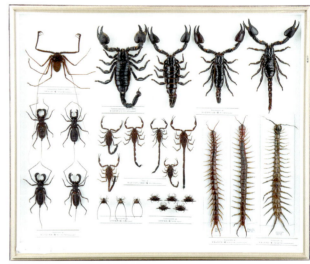

32

082　29. コーカサスオオカブト・アトラスオオカブト他［コガネムシ科］（鞘翅目）.
　　　　Chalcosoma chiron, Calcosoma atlas, etc. [Scarabaeidae] (Coleoptera).

　　30. ヒロキオビルリタマムシ［タマムシ科］・バイオリンムシ［オサムシ科］・ミド
　　　　リツヤダイコクコガネ［コガネムシ科］他（鞘翅目）.
　　　　Chrysochroa maruyamai [Buprestidae], *Mormolyce phyllodes* [Carabidae],
　　　　Oxysternon conspicillatum [Scarabaeidae], etc. (Coleoptera).

31. イチジクカミキリ・アオオビハデツヤカミキリ他［カミキリムシ科］（鞘翅目）.
　　Batocera rubus, Anoplophora medembachi, etc. [Cerambycidae] (Coleoptera).

32. グレイウデムシ［ウデムシ科］（ウデムシ目）・ダイオウサソリ［コガネサソリ
　　科］（サソリ目）他.
　　Charon grayi [Charontidae] (Amblypygi), *Pandinus imperator* [Scorpionidae]
　　(Scorpiones), etc.

1-32. 箱寸法（Box dimantions）418×507 mm.

須田孫七
―― 日本のファーブル

Magoshichi Suda
— Regarded as the "Japanese Fabre"

須田孫七（1931〜2018）

東京都杉並区に生まれた須田孫七（1931〜2018年）は、幼少期は病弱で外出もままならないほどであったため、庭先にいる昆虫をはじめとする身近な動植物に興味を持つようになる。孫のことを不憫に思った祖母が背負い、井の頭恩賜公園へ度々散歩に行ったことも、その後の自然や生きものに対する見識の基礎を養うこととなった。

小学校入学後には体調も徐々に回復し、先進的な理科教育を実践されていた西沢二郎教諭と出会い、昆虫採集や動植物の調査研究に取り組むようになる。当時、井の頭にあった平山博物館、石神井にあった蝉類博物館にも足繁く通い、その指導を受けたことも大きい。井の頭自然文化園の開設にあたり、加藤氏からの依頼で井の頭を中心とした地域での採集と標本作成を熱心に行う。東京学芸大学に進学後は古川晴男教授に師事し昆虫学を専攻。東京都の公立中学教諭となり、子供たちや一般市民を対象とした昆虫学の普及啓発や後進の育成、身近な自然を調べ守り残す取り組みに尽力した。

10万点以上からなる須田コレクションは、多様な分類群を含み、収集範囲も国内外の広範に及ぶ。特に注目されるのは1940年前後から2000年代にかけて収集された東京都産の標本群であり、東京の環境と昆虫相の変遷を知ることができるものとなっている。また、小型の昆虫、特にアリが好きだったため、その標本も充実している。

（須田真一）

Mr. Magoshichi Suda (1931 – 2018) was born in Suginami-ku, Tokyo. In his childhood, he became fascinated with the insects, animals and plants in the garden of his house because he was housebound due to illness. His grandmother, feeling sorry for him, took him on her back and often went for walks in Inokashira Park. This experience cemented his passion, and from that day onwards he cultivated the base of knowledge of nature and creatures.

His physical condition gradually recovered during his junior school days, and after fully recovering he met Mr. Jiro Nishizawa, a science teacher who kindled his passion for insect collection. At that time he frequented the Hirayama Museum (owned by Mr. Shujiro Hirayama) in Inokashira and the Cicada Museum (owned by Dr. Masayo Kato) in Shakujii, where he received tuition in entomology. When the Inokashira Park Zoo opened, he devoted himself to collecting insects in the Inokashira area after a request from Dr. Kato. While at Tokyo Gakugei University, he studied entomology under the guidance of Prof. Haruo Furukawa. After graduation, he taught at a public junior high school and was also involved in educational activities through entomology and nature research.

His collection, which is comprised of over 100,000 specimens, contains various insect taxa from a wide range of countries. Of particular note are many insect specimens from Tokyo collected in the 1940s–2000s. Using these specimens, we can understand the changes in local insect fauna and nature over that period. Moreover, numerous ant specimens are included in the collection as a result of his particular fascination in this group.

(Shin-ichi Suda)

1

2

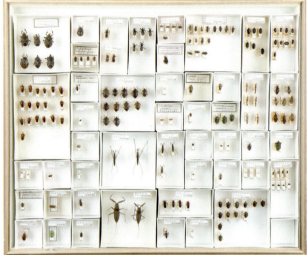

3

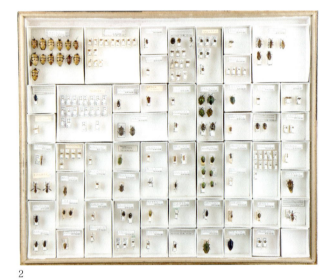

4

084 1. クロゴキブリ［ゴキブリ科］（蜚蠊目）・オオカマキリ［カマキリ科］（蟷螂目）
他.
Periplaneta fuliginosa [Blattellidae] (Blattaria), *Tenodera aridifolia* [Mantidae] (Mantodea), etc.

2. アカギカメムシ・アカスジキンカメムシ［キンカメムシ科］・チャバネアオカメムシ
［カメムシ科］他（半翅目）.
Cantao ocellatus, *Poecilocoris lewisi* [Scutelleridae], *Plautia stali* [Pentatomidae], etc. (Hemiptera).

3. エサキモンキツノカメムシ［ツノカメムシ科］・イトカメムシ［イトカメムシ
科］・オオアメンボ［アメンボ科］他（半翅目）.
Sastragala esakii [Acanthosomatidae], *Yemma exilis* [Berytidae], *Aquarius elongatus* [Gerridae], etc. (Hemiptera).

4. ハナアブ［ハナアブ科］・オオクロバエ［クロバエ科］・ビロウドツリアブ［ツリ
アブ科］他（双翅目）.
Eristalis tenax [Syrphidae], *Calliphora nigribarbis* [Calliphoridae], *Bombylius major* [Bombyliidae], etc. (Diptera).

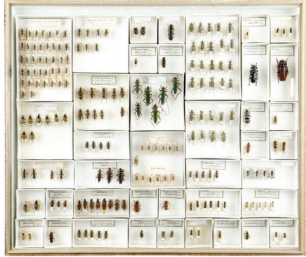

5. オオクワガタ・ミヤマクワガタ・ツヤハダクワガタ他［クワガタムシ科］（鞘翅目）.
   Dorcus hopei, Lucanus maculifemoratus, Ceruchus ligunarius, etc. [Lucanidae] (Coleoptera).

6. シカツノハナムグリ・クロカナブン・タイワンシロテンハナムグリ他［コガネムシ科］（鞘翅目）.
   Dicronocephalus wallichii, Rhomborhina polita, Protaetia orientalis sakaii, etc. [Scarabaeidae] (Coleoptera).

7. オオヨツスジハナカミキリ・ルリボシカミキリ・クスベニカミキリ他［カミキリムシ科］（鞘翅目）.
   Macroleptura regalis, Rosalia batesi, Pyrestes nipponicus, etc. [Cerambycidae] (Coleoptera).

8. オオルリハムシ・ルリハムシ・イタドリハムシ他［ハムシ科］（鞘翅目）.
   Chrysolina virgata, Plagiosterna aenea, Gallerucida bifasciata, etc. [Chrysomelidae] (Coleoptera).

9

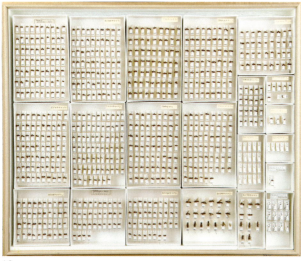
10

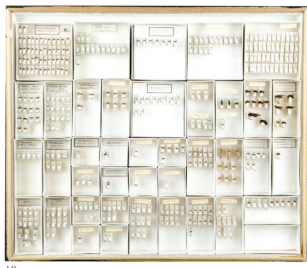
11

12

086

9. アリ類（膜翅目）.
    Ants [Formicidae] (Hymenoptera).

10. アリ類（膜翅目）.
    Ants [Formicidae] (Hymenoptera).

11. アリ類（膜翅目）.
    Ants [Formicidae] (Hymenoptera).

12. アリ類（膜翅目）.
    Ants [Formicidae] (Hymenoptera).

15

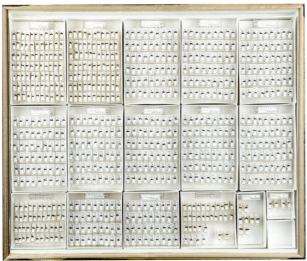

13

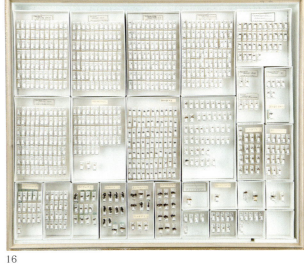

16

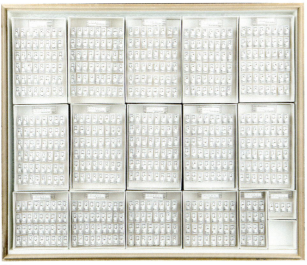

14

13. アリ類（膜翅目）.
　　Ants [Formicidae] (Hymenoptera).

14. アリ類（膜翅目）.
　　Ants [Formicidae] (Hymenoptera).

15. アリ類（膜翅目）.
　　Ants [Formicidae] (Hymenoptera).

16. アリ類（膜翅目）.
　　Ants [Formicidae] (Hymenoptera).

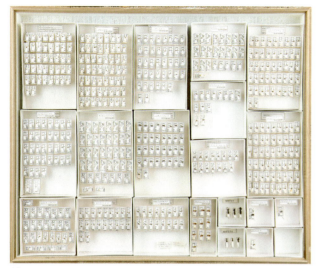
17

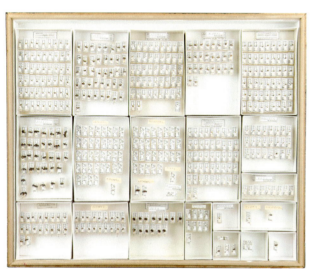
18

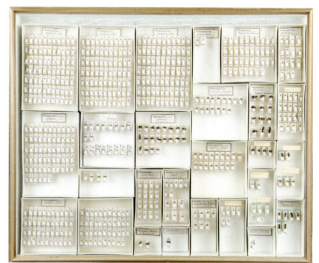
19

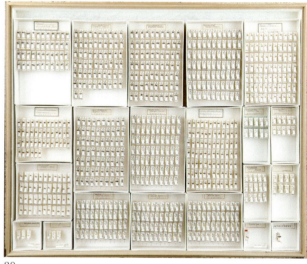
20

17. アリ類（膜翅目）.
    Ants [Formicidae] (Hymenoptera).

18. アリ類（膜翅目）.
    Ants [Formicidae] (Hymenoptera).

19. アリ類（膜翅目）.
    Ants [Formicidae] (Hymenoptera).

20. アリ類（膜翅目）.
    Ants [Formicidae] (Hymenoptera).

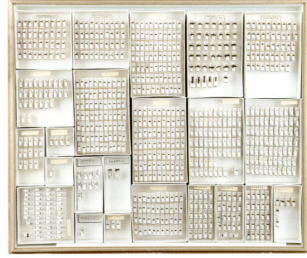
23

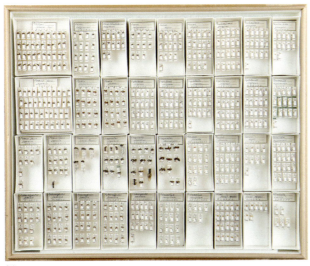
21

22

24

21. アリ類（膜翅目）.
    Ants [Formicidae] (Hymenoptera).

22. アリ類（膜翅目）.
    Ants [Formicidae] (Hymenoptera).

23. アリ類（膜翅目）.
    Ants [Formicidae] (Hymenoptera).

24. アリ類（膜翅目）.
    Ants [Formicidae] (Hymenoptera).

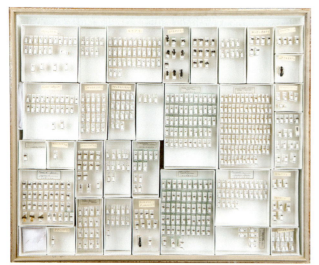
25

27

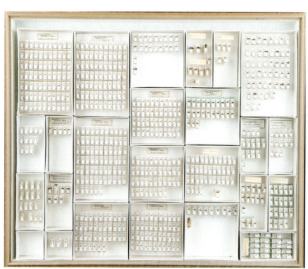
26

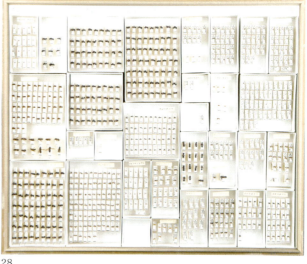
28

090

25. アリ類（膜翅目）．
　　Ants [Formicidae] (Hymenoptera).

26. アリ類（膜翅目）．
　　Ants [Formicidae] (Hymenoptera).

27. アリ類（膜翅目）．
　　Ants [Formicidae] (Hymenoptera).

28. アリ類（膜翅目）．
　　Ants [Formicidae] (Hymenoptera).

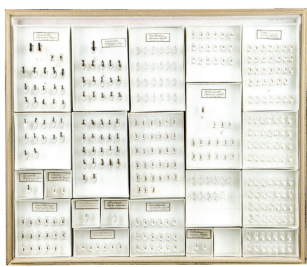
31

29

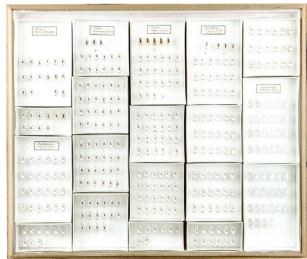
32

30

29. アリ類（膜翅目）.
   Ants [Formicidae] (Hymenoptera).

30. アリ類（膜翅目）.
   Ants [Formicidae] (Hymenoptera).

31. アリ類（膜翅目）.
   Ants [Formicidae] (Hymenoptera).

32. アリ類（膜翅目）.
   Ants [Formicidae] (Hymenoptera).

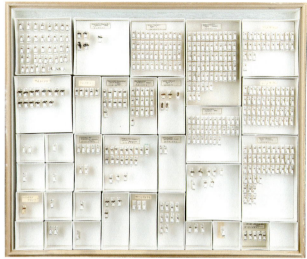

33.

34.

35.

36.

33. アリ類（膜翅目）.
   Ants [Formicidae] (Hymenoptera).

34. 東京の昆虫：キイトトンボ［イトトンボ科］・トラフトンボ［エゾトンボ科］・キトンボ［トンボ科］（蜻蛉目）他.
   Ceriagrion melanurum [Coenagrionidae], *Epitheca marginata* [Corduliidae], *Sympetrum croceolum* [Libellulidae] (Odonata), etc. from Tokyo.

35. 東京の昆虫：グンバイトンボ［モノサシトンボ科］・アオハダトンボ［カワトンボ科］（蜻蛉目）・ナベブタムシ［ナベブタムシ科］（半翅目）他.
   Platycnemis foliacea [Platycnemidinae], *Calopteryx japonica* [Calopterygidae] (Odonata), *Aphelocheirus vittatus* [Aphelochiridae] (Hemiptera), etc. from Tokyo.

36. 東京の昆虫：ヨツボシトンボ［トンボ科］（蜻蛉目）・ヤマトセンブリ［センブリ科］（脈翅目）・アサマイチモンジ［タテハチョウ科］（鱗翅目）他.
   Libellula quadrimaculata [Libellulidae] (Odonata), *Sialis yamatoensis* [Sialidae] (Neuroptera), *Limenitis glorifica* [Nymphalidae] (Lepidoptera), etc. from Tokyo.

39

37

40

38

37. 東京の昆虫：タガメ［コオイムシ科］（半翅目）・スジゲンゴロウ・ナミゲンゴロウ［ゲンゴロウ科］（鞘翅目）他．
    Lethocerus deyrollei [Belostomatidae] (Hemiptera), *Hydaticus satoi, Cybister japonicus* [Dytiscidae] (Coleoptera), etc. from Tokyo.

38. 東京の昆虫：ナミアゲハ［アゲハチョウ科］（鱗翅目）・アブラゼミ［セミ科］（半翅目）他．
    Papilio xuthus [Papilionidae] (Lepidoptera), *Graptopsaltria nigrofuscata* [Cicadidae] (Hemiptera), etc. from Tokyo.

39. 東京の昆虫：ケラ［ケラ科］（直翅目）・キボシカミキリ［カミキリムシ科］（鞘翅目）他．
    Gryllotalpa orientalis [Gryllotalpidae] (Orthoptera), *Psacothea hilaris* [Cerambycidae] (Coleoptera), etc. from Tokyo.

40. 東京の昆虫：セグロバッタ［イナゴ科］（直翅目）・オオセイボウ［セイボウ科］（膜翅目）他．
    Shirakiacris shirakii [Catantopidae] (Orthoptera), *Stilbum cyanurum* [Chrysididae] (Hymenoptera), etc. from Tokyo.

41

42

43

44

41. 東京の昆虫：ギフチョウ［アゲハチョウ科］・クロシジミ［シジミチョウ科］・オオムラサキ［タテハチョウ科］（鱗翅目）他．
  Luehdorfia japonica [Papilionidae], *Niphanda fusca* [Lycaenidae], *Sasakia charonda* [Nymphalidae], etc. from Tokyo.

42. 東京の昆虫：コツバメ［シジミチョウ科］・オナガアゲハ［アゲハチョウ科］・ヒメキマダラセセリ［セセリチョウ科］（鱗翅目）他．
  Callophrys ferrea [Lycaenidae], *Papilio macilentus* [Papilionidae], *Ochlodes ochraceus* [Hesperiidae] (Lepidoptera), etc. from Tokyo.

43. 東京の昆虫：カブトムシ［コガネムシ科］（鞘翅目）・クスサン［ヤママユガ科］（鱗翅目）・オオマルハナバチ［ミツバチ科］（膜翅目）他．
  Trypoxylus dichotomus [Scarabaeidae] (Coleoptera), *Caligula japonica* [Saturniidae] (Lepidoptera), *Bombus hypocrita* [Apidae] (Hymenoptera), etc. from Tokyo.

44. 東京の昆虫：ヤマトタマムシ［タマムシ科］（鞘翅目）・キカマキリモドキ［カマキリモドキ科］・ラクダムシ［ラクダムシ科］（脈翅目）他．
  Chrysochroa fulgidissima [Buprestidae] (Coleoptera), *Eumantispa harmandi* [Mantispidae], *Inocellia japonica* [Inocelliidae] (Neuroptera). etc. from Tokyo.

45

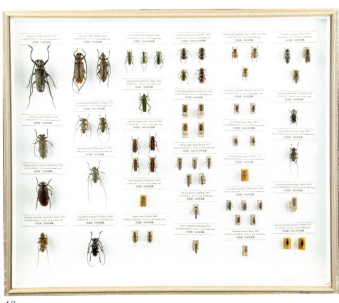
46

45. 東京の昆虫：カワラバッタ［バッタ科］（直翅目）・オオルリハムシ［ハムシ科］（鞘翅目）・ツマグロキチョウ［シロチョウ科］（鱗翅目）他.
Eusphingonotus japonicus [Acrididae] (Orthoptera), *Chrysolina virgata* [Chrysomelidae] (Coleoptera), *Eurema laeta* [Pieridae] (Lepidoptera), etc. from Tokyo.

46. 東京の昆虫：ハンノキカミキリ・シロスジカミキリ［カミキリムシ科］（鞘翅目）他.
Cagosima sanguinolenta, *Batocera lineolata* [Cerambycidae] (Coleoptera), etc. from Tokyo.

1-46. 箱寸法 (Box dimantions) 418×507 mm.

濱　正彦
——信州の蝶聖

Masahiko Hama
— Saint Aurelian of Shinshu

濱　正彦（1935〜2012）

長野県茅野市に生まれた濱 正彦（1935〜2012年）は、生涯を通して昆虫の研究に情熱を傾け、信州屈指のアマチュア研究者として名を馳せた。特に信州昆虫学会（編）による名著「信濃の蝶」シリーズのデータは同氏による功績がかなり大きい。1954年に長野県立諏訪清陵高校を卒業し、諏訪精工舎に入社、1960年頃塩尻工場設立に伴い塩尻に転居し、1995年に退職した。そのため、1950年代までは諏訪での蝶類生態研究に集中したが、1960年代以降は塩尻に拠点を移し、その後は車で調査できる範囲の長野県全体に活動を広げた。同時にこの頃からガ類やカミキリムシ類、オサムシ類にも採集の範囲を拡大している。

　濱コレクションの内容は、大型ドイツ箱294箱に収納されたチョウ類標本44,570頭で構成される。主として長野県中南部産のチョウ類で、1950年代から採集や飼育し続けたものである。希少種だけでなく、普通種までも丹念に集められ、長い時間をかけて収集された結果、生息の消長、地理的変異、個体変異等を調べる上での重要な情報を与える標本となっている。また、日本産チョウ類の比較形態学的研究や分子系統学研究等に、有力な原資料を提供するものである。彼のガ類コレクションは飯田市美術博物館に保管されているが、このチョウ類コレクションは、2013年4月に長男・濱 和彦氏から東京大学総合研究博物館に寄贈された。

(矢後勝也)

　Mr. Masahiko Hama (1935–2012), born in Chino City, Nagano, was fascinated with researching and collecting insects through his life and was also widely known as the foremost amateur entomologist in the Shinshu area. In particular, he contributed in a significant way to a series of great books "Butterflies of Shinano" edited by the Entomological Society of Shinshu. After graduating from Suwa Seiryo High School, he joined Suwa Seikosha Co., Ltd. After the company opened a factory in the Shiojiri region, he was transferred to Shiojiri City in 1960, where he worked until his retirement in 1995. Therefore, in the 1950s he investigated the biology of butterflies in Suwa City, and then based himself in Shiojiri City from the 1960s onward. Thanks to his access to a car, he was able to extend the range of his research to the entire Nagano region. He simultaneously collected moths and carabid beetles around this time.

　His butterfly collection contains 44,570 specimens in 294 specimen boxes, comprised mainly of butterflies from central and southern Nagano which he had collected and reared over his life since the 1950s. As he diligently collected not only rare species but also common ones over such a long period, the collection gives important information for research on life cycles, geographical variation and population variation. Moreover, these specimens provide us with useful information for morphological and molecular studies. Although his moth collection remains housed in the Iida City Museum, Nagano, his butterfly collection was donated by his eldest son, Mr. Kazuhiko Hama, to the University Museum, the University of Tokyo, in Apr. 2013.

(Masaya Yago)

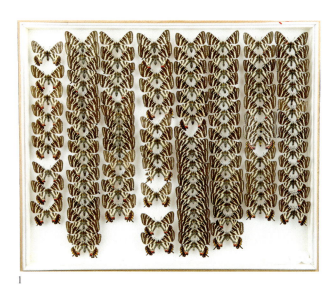

1

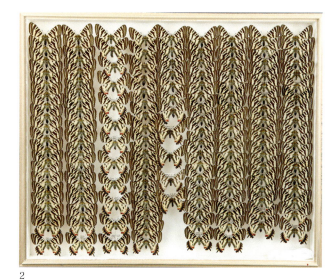

3

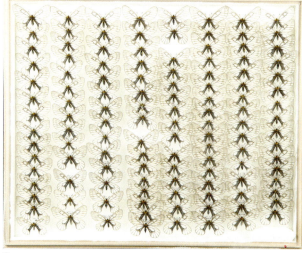

2

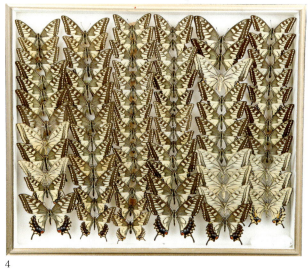

4

1. ギフチョウ［アゲハチョウ科］（鱗翅目）．
   Luehdorfia japonica [Papilionidae] (Lepidoptera).

2. ヒメギフチョウ［アゲハチョウ科］（鱗翅目）．
   Luehdorfia puziloi [Papilionidae] (Lepidoptera).

3. ウスバシロチョウ［アゲハチョウ科］（鱗翅目）．
   Parnassius citrinarius [Papilionidae] (Lepidoptera).

4. キアゲハ［アゲハチョウ科］（鱗翅目）．
   Papilio machaon [Papilionidae] (Lepidoptera).

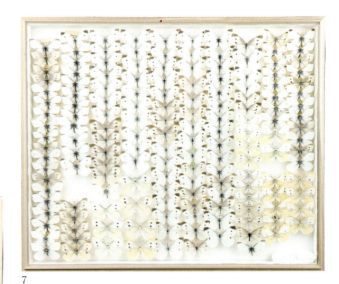

7

5

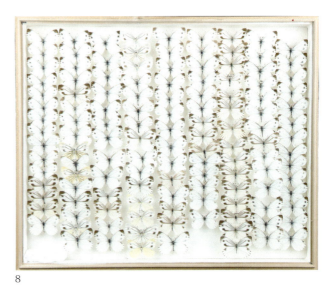

8

6

098

5. アゲハ・クロアゲハ［アゲハチョウ科］（鱗翅目）.
   Papilio xuthus & *P. protenor* [Papilionidae] (Lepidoptera).

6. ミヤマカラスアゲハ［アゲハチョウ科］（鱗翅目）.
   Papilio maackii [Papilionidae] (Lepidoptera).

7. モンシロチョウ［シロチョウ科］（鱗翅目）.
   Pieris rapae [Pieridae] (Lepidoptera).

8. ヤマトスジグロシロチョウ［シロチョウ科］（鱗翅目）.
   Pieris nesis [Pieridae] (Lepidoptera).

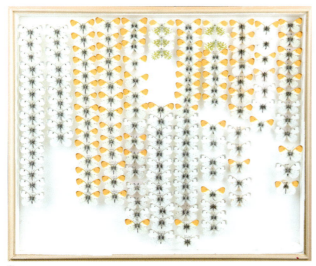

9

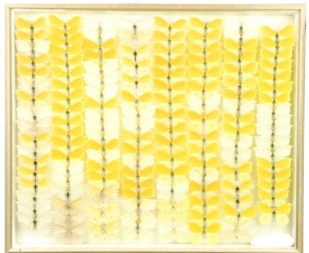

11

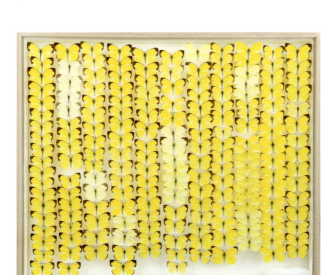

10

12

9. クモマツマキチョウ［シロチョウ科］（鱗翅目）.
   Anthocharis cardamines [Pieridae] (Lepidoptera).

10. キタキチョウ［シロチョウ科］（鱗翅目）.
    Eurema mandarina [Pieridae] (Lepidoptera).

11. スジボソヤマキチョウ［シロチョウ科］（鱗翅目）.
    Gonepteryx aspasia [Pieridae] (Lepidoptera).

12. ウラギンシジミ・ムラサキシジミ［シジミチョウ科］（鱗翅目）.
    Curetis acuta & *Arhopala japonica* [Lycaenidae] (Lepidoptera).

15

13

16

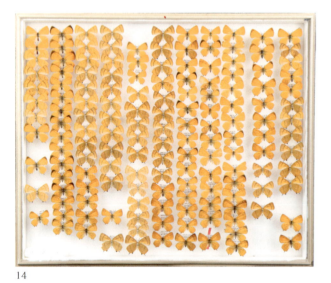

14

100

13. ムラサキツバメ・ウラキンシジミ［シジミチョウ科］（鱗翅目）．
    Arhopala bazalus & *Ussuriana stygiana* [Lycaenidae] (Lepidoptera).

14. ムモンアカシジミ［シジミチョウ科］（鱗翅目）．
    Shirozua jonasi [Lycaenidae] (Lepidoptera).

15. フジミドリシジミ・ウラゴマダラシジミ［シジミチョウ科］（鱗翅目）．
    Sibataniozephyrus fujisanus & *Artopoetes pryeri* [Lycaenidae] (Lepidoptera).

16. アイノミドリシジミ［シジミチョウ科］（鱗翅目）．
    Chrysozephyrus brillantinus [Lycaenidae] (Lepidoptera).

17

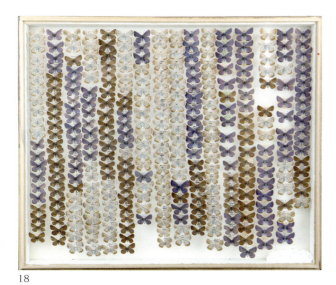

18

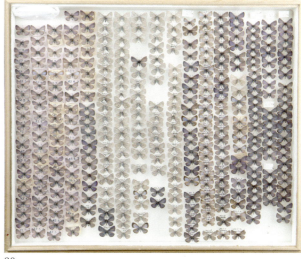

19

20

101

17. ゴマシジミ［シジミチョウ科］（鱗翅目）．
    Phengaris teleius [Lycaenidae] (Lepidoptera).

18. ミヤマシジミ［シジミチョウ科］（鱗翅目）．
    Plebejus argyrognomon [Lycaenidae] (Lepidoptera).

19. アサマシジミ［シジミチョウ科］（鱗翅目）．
    Plebejus subsolanus [Lycaenidae] (Lepidoptera).

20. スギタニルリシジミ［シジミチョウ科］（鱗翅目）．
    Celastrina sugitanii [Lycaenidae] (Lepidoptera).

23

21

24

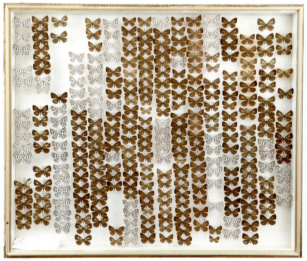

22

102

21. クロツバメシジミ[シジミチョウ科]（鱗翅目）.
    Tongeia fischeri [Lycaenidae] (Lepidoptera).

22. ゴイシシジミ[シジミチョウ科]（鱗翅目）.
    Taraka hamada [Lycaenidae] (Lepidoptera).

23. ベニシジミ[シジミチョウ科]（鱗翅目）.
    Lycaena phlaeas [Lycaenidae] (Lepidoptera).

24. クジャクチョウ[タテハチョウ科]（鱗翅目）.
    Inachis io [Nymphalidae] (Lepidoptera).

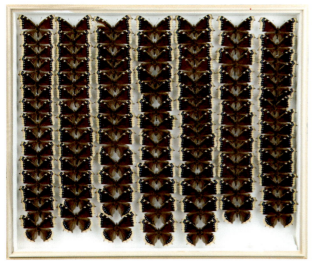

25

26

27

28

25. キベリタテハ［タテハチョウ科］（鱗翅目）.
    Nymphalis antiopa [Nymphalidae] (Lepidoptera).

26. エルタテハ［タテハチョウ科］（鱗翅目）.
    Nymphalis vaualbum [Nymphalidae] (Lepidoptera).

27. サカハチチョウ・コヒョウモンモドキ［タテハチョウ科］他（鱗翅目）.
    Araschnia burejana, Melitaea ambigua [Nymphalidae], etc. (Lepidoptera).

28. ミスジチョウ［タテハチョウ科］・アオバセセリ［セセリチョウ科］他（鱗翅目）.
    Neptis philyra [Pieridae], *Choaspes benjaminii* [Nymphalidae], etc. (Lepidoptera).

29

30

31

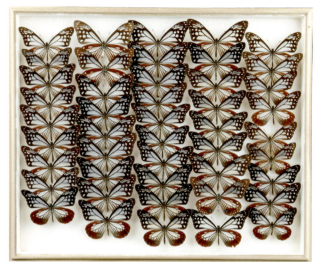

32

33

29. スミナガシ [タテハチョウ科] (鱗翅目).
    Dichorragia nesimachus [Nymphalidae] (Lepidoptera).

30. コムラサキ [タテハチョウ科] (鱗翅目).
    Apatura metis [Nymphalidae] (Lepidoptera).

31. オオムラサキ・オオウラギンスジヒョウモン [タテハチョウ科] (鱗翅目).
    Sasakia charonda & *Argyronome ruslana* [Nymphalidae] (Lepidoptera).

32. アサギマダラ [タテハチョウ科] (鱗翅目).
    Parantica sita [Nymphalidae] (Lepidoptera).

33. ヒメヒカゲ [タテハチョウ科] (鱗翅目).
    Coenonympha oedippus [Nymphalidae] (Lepidoptera).

1-33. 箱寸法 (Box dimantions) 418×507 mm.

宮野浩二
—— 西多摩の博物学者

Koji Miyano
— Great naturalist from Nishitama, Tokyo

宮野浩二（1937〜2012）

　宮野浩二（1937〜2012年）は、東京都大久野村（現・日の出町）に生まれた博物学者である。東京理科大学理学部を卒業後、1961年に第一ゴム製造KK入社、1964年からは都内の公共職業安定所に勤務された。古くから標本作りを行っていたが、1960年代半ばから動植物に衰退の兆候があるのを察知し、動植物を採集して標本を後世に残すことに執着するきっかけとなった。特にチョウや甲虫のような人気のある昆虫だけでなく、昆虫全般を扱って収集した。また、動植物全般に広い見識を持ち、発足メンバーの一人として地元の自然科学同好会「幸神蝶類同好会（1966）」や後続の同好会「日の出自然科学同好会（1967〜1982）」を立ち上げて、子供達に科学する心を芽生えさせることに尽力した。近所の子供達が持ってきた昆虫も丹念に標本にして、ラベルを付けて残している。こうして構築された昆虫コレクションは膨大で、特に1960年代から1970年代の標本が多い。

　宮野コレクションは昆虫だけでなく、クモ、鳥、哺乳類、植物、化石、岩石など様々な自然物から構成され、その総数は3万点以上に及ぶ。このうち昆虫標本は約2万頭と大半を占め、現在の東京都では絶滅した可能性が高いアカセセリやホシチャバネセセリ、ヒメシロチョウ、タガメ、ゲンゴロウなど貴重な種の標本が数多く含まれる。日の出ネイチャークラブの会員らによりこれらの標本に関する目録も作成されている。本コレクションは、2017年11月に奥方・伸子氏から東京大学総合研究博物館に寄贈された。

（矢後勝也）

　Mr. Koji Miyano (1937–2012) was a local museologist born in Oguno-mura (now Hinode-machi), Tokyo. After graduation from the Faculty of Science, Tokyo University of Science, he joined "Daiichi Gum Seizo KK" in 1961, and made a career move to the Tokyo public employment security service in 1964, where he worked until his retirement. He was fascinated with insects, animals and plants, collecting and preserving them since his youth. From the mid 1960s, he noticed that the number of living things in his neighborhood was gradually decreasing, so he had a strong desire to collect them to preserve for future generations. He not only collected and stored butterflies and beetles but also other less popular insects. Possessing a wide range of knowledge in animals and plants, he co-established a local science club "The Sajigami Butterfly Club (1966)" and its successor "The Hinode Science Club (1967-1982)" which were involved in science education through nature research. He treated even the insects brought by neighboring children with the same respect as his own samples, mounting them carefully and storing them with labels in specimen boxes. The insect collection constructed in this process is huge, notably with many specimens collected in the 1960s-1970s.

　His collection is composed of not only various insects but also spiders, birds, animals, plants and fossils. The total number reaches more than 30,000 specimens, with insect specimens accounting for the majority, including many species faced with extinction from the Tokyo area such as *Hesperia florinda, Aeromachus inachus, Leptidea amurensis, Lethocerus deyrollei* and *Cybister japonicus*. The catalogue of his insect collection was published by members of the Hinode Nature Club in 2015. The donation of his collection to the University Museum, The University of Tokyo was carried out according to the wishes of his wife, Mrs. Nobuko Miyano, in Nov. 2017.

(Masaya Yago)

1

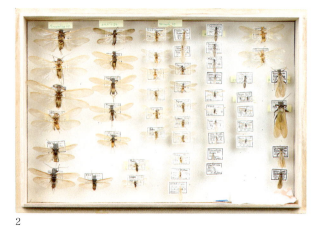
2

3

4

1. イシノミ類(古顎目)・トビムシ類(粘管目)・キジラミ類(半翅目).
   Rock bristletails (Archeognatha), springtails (Collembola) & psyllas (Hemiptera).
2. カワゲラ類(襀翅目).
   Stoneflies (Plecoptera).
3. ハサミムシ類(革翅目).
   Earwigs (Dermaptera).
4. コオロギ・バッタ類(直翅目).
   Crickets & grasshoppers (Orthoptera).

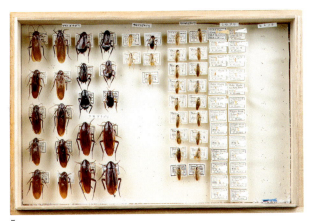
5

6

7

8

5. ゴキブリ類(蜚蠊目)・シロアリ類(等翅目).
   Cockroaches (Blattodea) & termites (Isoptera).

6. ハゴロモ類・ウンカ類・グンバイ類(半翅目).
   Planthoppers & lace-bugs (Hemiptera).

7. カメムシ類(半翅目).
   Stink bugs (Hemiptera).

8. カメムシ・アメンボ類(半翅目).
   Stink bugs & water striders (Hemiptera).

9

10

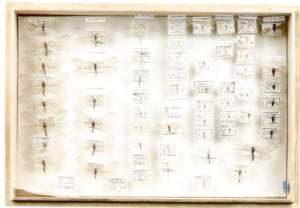
11

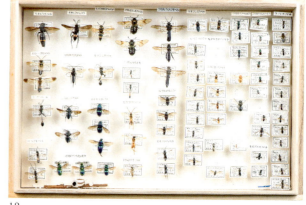
12

9. タガメ・タイコウチ・ミズカマキリ・マツモムシ類(半翅目).
   Giant warter bugs, water stick insects, backswimmers, etc. (Hemiptera).

10. チャタテムシ類(咀顎目)・ラクダムシ(駱駝虫目)・アブラムシ類(半翅目)他.
    Psocids (Psocodea), snakefly (Raphidioptera), aphids (Hemiptera), etc.

11. アミメカゲロウ類(脈翅目).
    Lacewings & alderflies (Neuroptera).

12. ハバチ類・セイボウ類・ハナバチ類他(膜翅目).
    Sawflies, jewel wasps, bees, etc. (Hymenoptera).

13

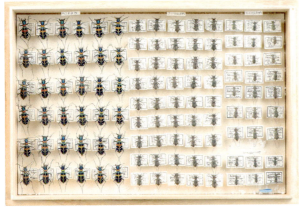

14

15

16

13. アリバチ類・ジガバチ類・ヒメバチ類他（膜翅目）.
    Velvet ants, mud daubers, ichneumon flies, etc. (Hymenoptera).

14. ハンミョウ科（鞘翅目）.
    Tiger beetles [Cicindelidae] (Coleoptera).

15. ハンミョウ科（鞘翅目）.
    Tiger beetles [Cicindelidae] (Coleoptera).

16. マイマイカブリ［オサムシ科］（鞘翅目）.
    Damaster blaptoides [Carabidae] (Coleoptera).

17

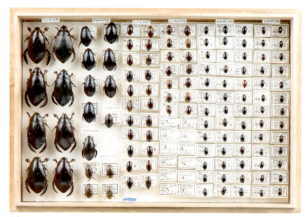

18

19

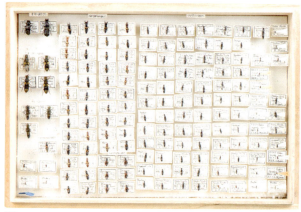

20

17. アオオサムシ［オサムシ科］（鞘翅目）.
    Carabus insulicola [Carabidae] (Coleoptera).

18. ナミゲンゴロウ・コガタノゲンゴロウ他［ゲンゴロウ科］（鞘翅目）.
    Cybister japonicus, C. tripunctatus, etc. [Dytiscidae] (Coleoptera).

19. ガムシ科（鞘翅目）.
    Water-scavenger beetles [Hydrophilidae] (Coleoptera).

20. ハネカクシ科（鞘翅目）.
    Rove beetles [Staphylinidae] (Coleoptera).

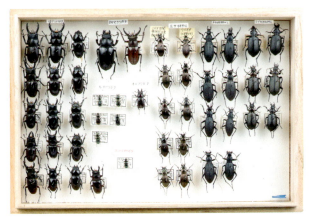
21

22

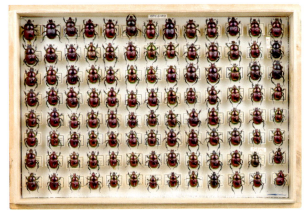
23

24

21. クワガタムシ科・オサムシ科（鞘翅目）.
    Stag beetles [Lucanidae] & ground beetles [Carabidae] (Coleoptera).

22. ミヤマクワガタ［クワガタムシ科］（鞘翅目）.
    Lucanus maculifemoratus [Lucanidae] (Coleoptera).

23. オオセンチコガネ・センチコガネ［センチコガネ科］（鞘翅目）.
    Geotrupes auratus & *G. laevistriatus* [Geotrupidae] (Coleoptera).

24. センチコガネ［センチコガネ科］・コカブトムシ・オオスジコガネ［コガネムシ科］, etc.（鞘翅目）.
    Geotrupes laevistriatus [Geotrupidae], *Eophileurus chinensis* & *Mimela costata* [Scarabaeidae], etc. (Coleoptera).

25

26

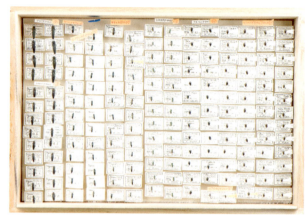

27

28

25. ハナムグリ類［コガネムシ科］（鞘翅目）.
　　Flower chafers [Scarabaeidae] (Coleoptera).

26. コガネムシ類・ハナムグリ類［コガネムシ科］（鞘翅目）.
　　Scarab beetles & flower chafers [Scarabaeidae] (Coleoptera).

27. タマムシ科（鞘翅目）.
　　Jewel beetles [Buprestidae] (Coleoptera).

28. タマムシ科（鞘翅目）.
　　Jewel beetles [Buprestidae] (Coleoptera).

29

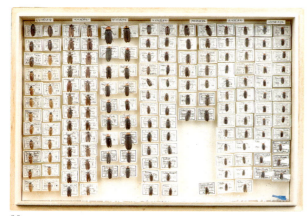

30

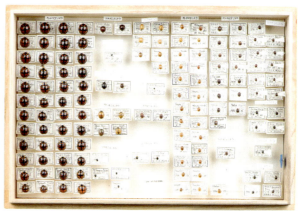

31

32

29. ヤマトタマムシ［タマムシ科］（鞘翅目）.
    Chrysochroa fulgidissima [Buprestidae] (Coleoptera).

30. ホタル科（鞘翅目）.
    Fireflies [Lampyridae] (Coleoptera).

31. テントウムシ科（鞘翅目）.
    Ladybirds [Coccinellidae] (Coleoptera).

32. ゴミムシダマシ科（鞘翅目）.
    Darkling beetles [Tenebrionidae] (Coleoptera).

33

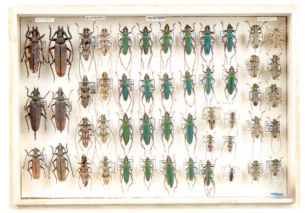
34

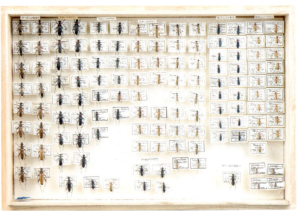
35

36

114

33. ツチハンミョウ科・クビナガムシ科・ハムシ科・タマムシ科・カミキリムシ科他（鞘翅目）.
    Meloidae, Stenotrachelidae, Chrysomelidae, Buprestidae, Cerambycidae, etc. (Coleoptera).

34. カミキリムシ科（鞘翅目）.
    Longicorn beetles [Cerambycidae] (Coleoptera).

35. カミキリムシ科（鞘翅目）.
    Longicorn beetles [Cerambycidae] (Coleoptera).

36. カミキリムシ科（鞘翅目）.
    Longicorn beetles [Cerambycidae] (Coleoptera).

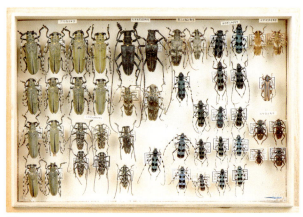
37

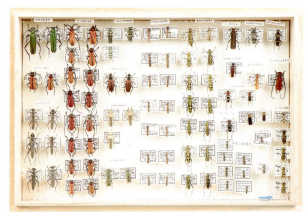
38

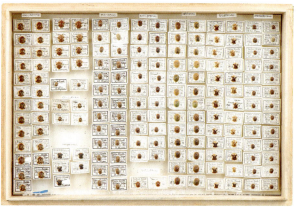
39

40

37. カミキリムシ科（鞘翅目）.
    Longicorn beetles [Cerambycidae] (Coleoptera).

38. カミキリムシ科（鞘翅目）.
    Longicorn beetles [Cerambycidae] (Coleoptera).

39. ハムシ科（鞘翅目）.
    Leaf beetles [Chrysomelidae] (Coleoptera).

40. ハムシ科（鞘翅目）.
    Leaf beetles [Chrysomelidae] (Coleoptera).

41

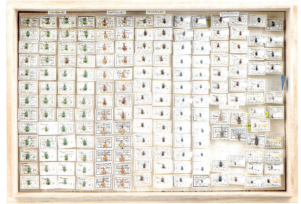

42

43

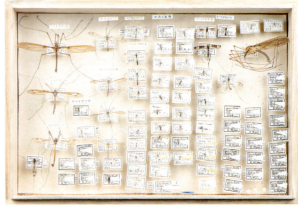

44

45

41. ゾウムシ科（鞘翅目）.
    Weevils [Curculionidae] (Coleoptera).

42. オトシブミ科（鞘翅目）.
    Leaf rolling weevils [Attelabidae] (Coleoptera).

43. 東京産チョウ類：ヒメシロチョウ[シロチョウ科]・ホシチャバネセセリ・アカ
    セセリ[セセリチョウ科] 他（鱗翅目）.
    Leptidea amurensis [Pieridae], *Aeromachus inachus*, *Hesperia florinda* [Hesperiidae], etc. (Lepidoptera) from Tokyo.

44. カ科・ガガンボ科（双翅目）.
    Mosquitos [Culicidae] & crane flies [Tipulidae] (Diptera).

45. アブ類（双翅目）.
    Horseflies [Tabanidae] & hoverflies [Syrphidae] (Diptera).

1-45. 箱寸法（Box dimantions）206×306 mm.

白石浩次郎
――平和・トンボ資料館館長

Kojiro Shiraishi
— Director of the Peace and Dragonfly Museum

白石浩次郎（1941～2010）
（柿沼　隆撮影）

　白石浩次郎氏（1941～2010年）は東京生まれ。幼少時より昆虫に興味を持ち、高校生の頃よりトンボの調査研究を本格的に始める。学生時代より国立予防衛生研究所の朝比奈正二郎博士の指導を受け、1957年設立の日本蜻蛉同好会（現・日本トンボ学会）の創設時会員でもある。コレクションの収集範囲は日本全国におよび、日本産トンボ目のほぼ全種をご自身で採集された。ただし、自然保護の観点から採集は必要最低限に留めていたために、標本点数は必ずしも多くない。普及啓発活動にも熱心に取り組まれ、1983年には石川一、新井裕、松木和雄と共に発起人となり、関東在住のトンボ研究者・同好者同士が情報交換等を行える場として「関東地方蜻蛉懇談会」を設立し、世話人として会場確保などを一貫して担当された。

　1989年には、東京都豊島区池袋の自宅2階を改装して「平和・トンボ資料館」を開設し、毎月2回一般に無料で開放していた。資料館の開設は奥方・寿美子氏の発案によるもので、開館前日には朝比奈博士が招かれて展示監修を行ったという。また、ご自宅に程近い区立池袋の森でトンボ類のモニリングを継続され、生息に大きな負の影響を与えている侵略的外来種アメリカザリガニの駆除など、保全活動にも尽力された。1950～1960年代の東京・埼玉や1980年代の小笠原で収集された標本資料は学術的な価値だけでなく、研究史の資料としても極めて重要である。コレクションは奥方を通じて東京大学総合研究博物館へ2010年12月に寄贈された。

（須田真一）

　Mr. Kojiro Shiraishi (1941–2010) was born in Tokyo. By his early childhood he had become enamored with the world of insects, and started researching dragonflies whilst in high school. During his school days he made some excellent achievements, studying under Dr. Shoziro Asahina (National Institute of Health), who was an authority on Odonata, and also helping found the Japanese Society for Odonatology, which was established in 1957. The dragonfly specimens in his collection were collected from all over Japan, including almost all of the Japanese dragonfly species and subspecies. A strong advocate for conservation, Shiraishi restricted his collecting to the minimum number of specimens necessary. In 1983, he established, in conjunction with Hajime Ishikara, Yutaka Arai and Kazuo Matsuki, "The Dragonfly Council of Kanto District" to promote dragonfly research and information exchange among researchers and enthusiasts.

　In 1989, he renovated the second floor of his house in Ikebukuro, Tokyo, to construct the Peace and Dragonfly Museum, which was open for free admission two days a month. It is said that the establishing of this museum was instigated by his wife, Mrs. Sumiko Shiraishi, and that the exhibition was supervised by Dr Asahina in the reception of this museum. He also monitored dragonflies at the "Ikebukuro-no-mori" municipal park near his home and conducted conservation activities such as the eradication of red swamp crayfish, an invasive alien species. Of his collection, the dragonfly specimens collected from Tokyo and Saitama in 1960s-1950s and from Ogasawara Islands in 1980s comprise an extremely valuable resource for taxonomic and historical studies. The donation of his collection to The University of Tokyo was made possible by his wife on December, 2010.

(Shin-ichi Suda)

1

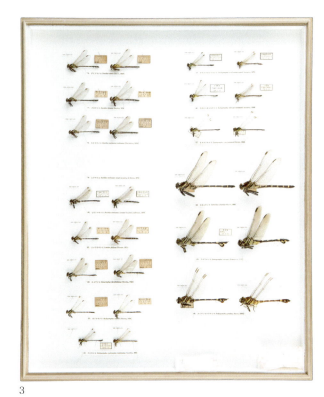

3

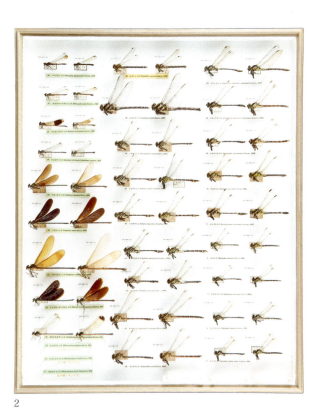

2

4

1. イトトンボ科・モノサシトンボ科・アオイトンボ科他［均翅亜目］（蜻蛉目）．
   Agrionidae, Platycnemididae, Lestidae, etc. [Zygoptera] (Odonata).

2. ハナダカトンボ科・カワトンボ科［均翅亜目］・ムカシトンボ科［均翅不均翅亜目］・サナエトンボ科・ムカシヤンマ科［不均翅亜目］（蜻蛉目）．
   Chlorocyphidae & Calopterygidae [Zygoptera], Epiophlebiidae [Anisozygoptera], Gomphidae & Petaluridae [Anisoptera] (Odonata).

3. サナエトンボ科［不均翅亜目］（蜻蛉目）．
   Gomphidae [Anisoptera] (Odonata).

4. オニヤンマ科・ミナミヤンマ科・ヤンマ科［不均翅亜目］（蜻蛉目）．
   Cordulegastridae, Chlorogomphidae & Aeshnidae [Anisoptera] (Odonata).

5

7

6

8

5. ヤンマ科 [不均翅亜目] (蜻蛉目).
   Aeshnidae [Anisoptera] (Odonata).

6. エゾトンボ科 [不均翅亜目] (蜻蛉目).
   Corduliidae [Anisoptera] (Odonata).

7. エゾトンボ科・トンボ科 [不均翅亜目] (蜻蛉目).
   Corduliidae & Libellulidae [Anisoptera] (Odonata).

8. トンボ科 [不均翅亜目] (蜻蛉目).
   Libellulidae [Anisoptera] (Odonata).

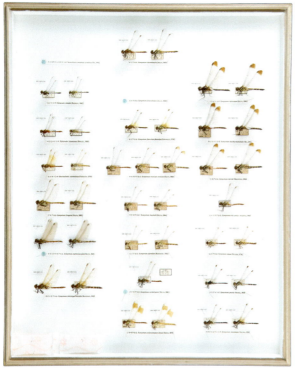

9

10

120

9. トンボ科［不均翅亜目］（蜻蛉目）.
   Libellulidae [Anisoptera] (Odonata).

10. トンボ科［不均翅亜目］（蜻蛉目）.
    Libellulidae [Anisoptera] (Odonata).

1-10. 箱寸法（Box dimantions）507×418 mm.

石川良輔
—— ハチ・オサムシ研究の巨匠

Ryôsuke Ishikawa
— Great entomologist studying bees and carabid beetles

石川良輔博士（1931年～）はハチ類とオサムシ類の研究で高名な昆虫学者である。京都府生まれで、横浜国立大学を卒業後、九州大学大学院農学研究科に進み、昆虫学を学んだ。1971年に北海道大学から理学博士取得した。国立科学博物館研究員を経て、東京都立大学理学部教授となる。1995年に退官し、現在は東京都立大学名誉教授。

研究の初期にハチ類の系統分類学的研究に携わり、やがてオサムシの系統分類や種分化の研究に邁進した。多くの論文や総説、図鑑を執筆し、中でも1991年出版の「オサムシを分ける錠と鍵」（八坂書房）は稀代の名著である。また、同博士の著作は昆虫だけに留まらず、自宅でカメとの日々の生活を綴った「うちのカメ オサムシの先生カメと暮らす」（八坂書房）は高い人気を博し、初版から25年となる現在でも販売されるほどのロングセラーを記録している。

石川コレクションはタイプ標本を含む多数のハチ類とオサムシ類の標本が主体で、インロウ箱約1,000箱に収納されている。このうちの一部が今夏に東京大学総合研究博物館へ寄贈され、今回の展示が実現した。残りの標本も今秋に本館へ収納される予定である。

（矢後勝也）

石川良輔（1931～）

Dr. Ryôsuke Ishikawa (1931 –), born in Kyoto, is a renowned entomologist researching bees and carabid beetles. After graduation from Yokohama National University, he went on to study the taxonomy of bees at the Faculty of Agriculture, Kyushu University. In 1971, he received a doctorate of science from Hokkaido University. Meanwhile, he held a position in the National Museum of Nature and Science, Tokyo, and later became a professor of Tokyo Metropolitan University. He retired in 1995, but continues as a professor emeritus of the university.

He first engaged in systematic studies of bees, and subsequently intently studied systematics and speciation of carabid beetles as a pioneer of this field. Moreover, he has published numerous articles, reviews and illustrated books including the widely-regarded titles "The Evolution of Carabus: Divergence and Isolating Mechanisms" (Yasaka Shobo). In addition, his published book "My Turtle and I –The Life and Behaviours of a Reeve's Turtle" (Yasaka Shobo) is still sold today as a long-selling book 25 years after its publication.

His insect collection, which comprises roughly 1,000 specimen boxes, is composed mainly of bees and carabid beetles. This summer, a part of the collection was kindly donated to the University Museum, the University of Tokyo, and for the first time is being exhibited here. The reminder of the collection will be also donated to the museum this autumn.

(Masaya Yago)

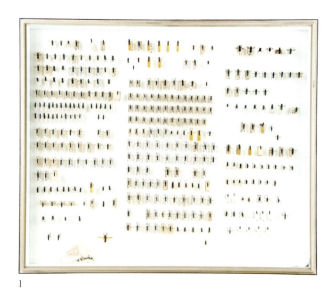
1

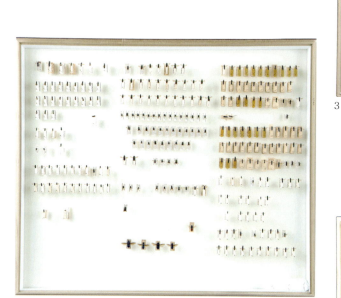
2

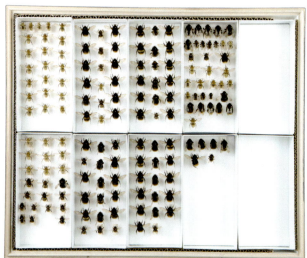
3

4

1. セイボウ類他 [セイボウ科] (膜翅目).
   Cockoo wasps [Chrysididae] (Hymenoptera).

2. セイボウ類他 [セイボウ科] (膜翅目).
   Cockoo wasps [Chrysididae] (Hymenoptera).

3. コマルハナバチ [ミツバチ科] (膜翅目).
   Bombus ardens [Apidae] (Hymenoptera).

4. ヒメマルハナバチ [ミツバチ科] (膜翅目).
   Bombus beaticola [Apidae] (Hymenoptera).

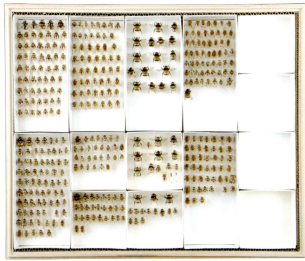

5

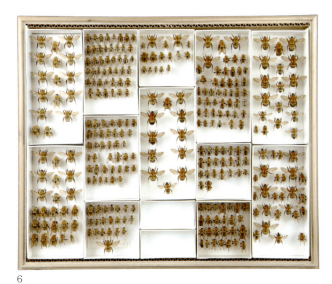

6

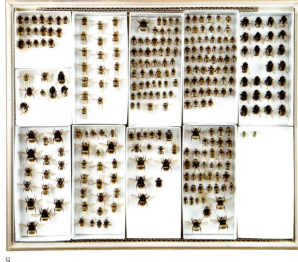

7

8

5. ヒメマルハナバチ［ミツバチ科］(膜翅目).
   Bombus beaticola [Apidae] (Hymenoptera).

6. トラマルハナバチ［ミツバチ科］(膜翅目).
   Bombus diversus [Apidae] (Hymenoptera).

7. ノサップマルハナバチ［ミツバチ科］他(膜翅目).
   Bombus florilegus [Apidae], etc. (Hymenoptera).

8. オオマルハナバチ［ミツバチ科］(膜翅目).
   Bombus hypocrita [Apidae] (Hymenoptera).

9

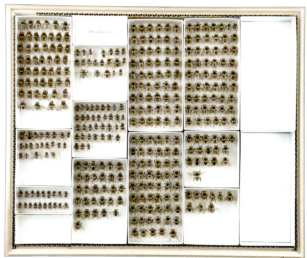
11

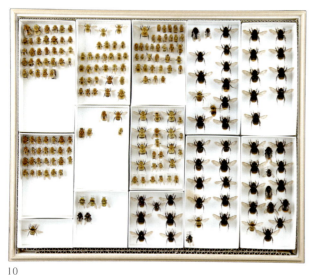
10

12

124

9. オオマルハナバチ [ミツバチ科] (膜翅目).
   Bombus hypocrita [Apidae] (Hymenoptera).

10. アカマルハナバチ・クロマルハナバチ [ミツバチ科] (膜翅目).
    Bombus hypnorum & *B. ignitus* [Apidae] (Hymenoptera).

11. ニセハイイロマルハナバチ [ミツバチ科] (膜翅目).
    Bombus pseudobaicalensis [Apidae] (Hymenoptera).

12. エゾナガマルハナバチ・ハイイロマルハナバチ他 [ミツバチ科] (膜翅目).
    Bombus yezoensis, *B. deuteronymus*, etc. [Apidae] (Hymenoptera).

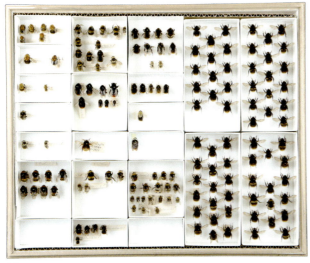

13

14

15

16

13. マルハナバチ類［ミツバチ科］（膜翅目）.
    Bumblebees [Apidae] (Hymenoptera).

14. マルハナバチ類［ミツバチ科］（膜翅目）.
    Bumblebees [Apidae] (Hymenoptera).

15. オサムシ科タイプ標本（鞘翅目）.
    Type specimens of carabid beetles [Carabidae] (Coleoptera).

16. オサムシの進化［オサムシ科］（鞘翅目）.
    Evolution of *Carabus* spp. [Carabidae] (Coleoptera).

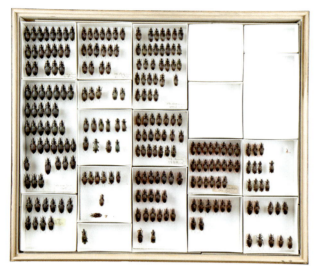
17

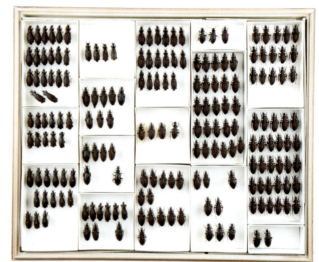
19

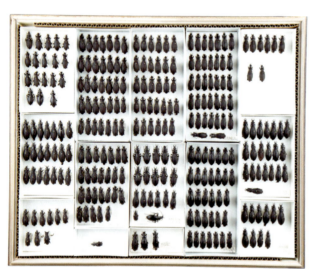
18

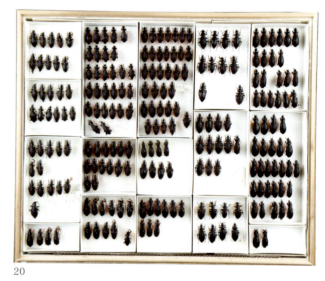
20

126

17. クロオサムシ・ルイスオサムシ［オサムシ科］（鞘翅目）.
    Carabus albrechti & *C. lewisianus* [Carabidae] (Coleoptera).

18. クロナガオサムシ［オサムシ科］（鞘翅目）.
    Carabus procerulus [Carabidae] (Coleoptera).

19. クロナガオサムシ［オサムシ科］（鞘翅目）.
    Carabus procerulus [Carabidae] (Coleoptera).

20. クロナガオサムシ［オサムシ科］（鞘翅目）.
    Carabus procerulus [Carabidae] (Coleoptera).

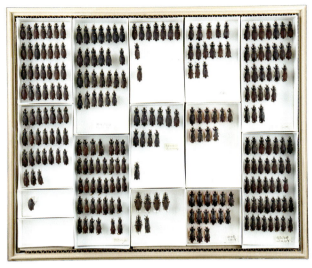

21

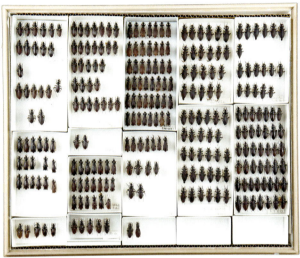

23

22

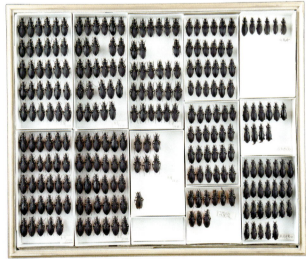

24

21. クロナガオサムシ［オサムシ科］（鞘翅目）.
    Carabus procerulus [Carabidae] (Coleoptera).

22. クロナガオサムシ［オサムシ科］（鞘翅目）.
    Carabus procerulus [Carabidae] (Coleoptera).

23. クロナガオサムシ［オサムシ科］（鞘翅目）.
    Carabus procerulus [Carabidae] (Coleoptera).

24. クロナガオサムシ［オサムシ科］（鞘翅目）.
    Carabus procerulus [Carabidae] (Coleoptera).

25

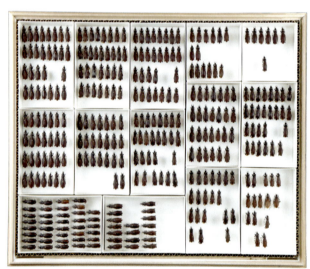

26

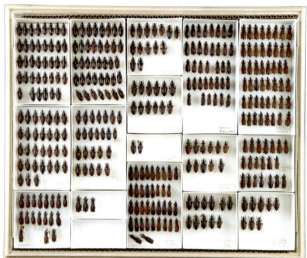

27

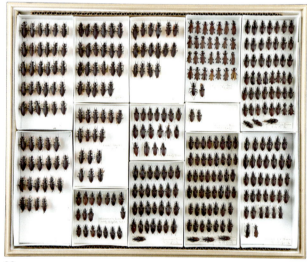

28

25. クロナガオサムシ［オサムシ科］（鞘翅目）.
   Carabus procerulus [Carabidae] (Coleoptera).

26. コクロナガオサムシ［オサムシ科］（鞘翅目）.
   Carabus arboreus [Carabidae] (Coleoptera).

27. コクロナガオサムシ［オサムシ科］（鞘翅目）.
   Carabus arboreus [Carabidae] (Coleoptera).

28. コクロナガオサムシ［オサムシ科］（鞘翅目）.
   Carabus arboreus [Carabidae] (Coleoptera).

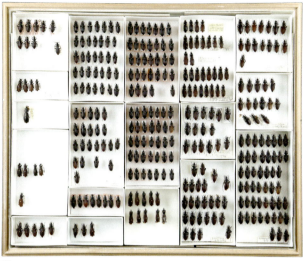

29

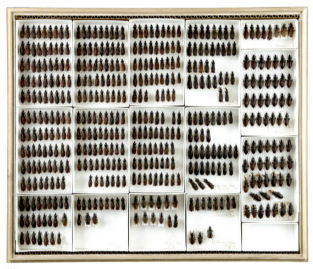

30

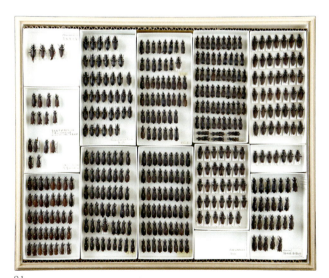

31

32

33

29. コクロナガオサムシ［オサムシ科］（鞘翅目）.
   Carabus arboreus [Carabidae] (Coleoptera).

30. コクロナガオサムシ［オサムシ科］（鞘翅目）.
   Carabus arboreus [Carabidae] (Coleoptera).

31. コクロナガオサムシ［オサムシ科］（鞘翅目）.
   Carabus arboreus [Carabidae] (Coleoptera).

32. コクロナガオサムシ［オサムシ科］（鞘翅目）.
   Carabus arboreus [Carabidae] (Coleoptera).

33. コクロナガオサムシ［オサムシ科］（鞘翅目）.
   Carabus arboreus [Carabidae] (Coleoptera).

1-33. 箱寸法（Box dimantions）418×507 mm.

尾本惠市
——二刀流の東京大学名誉教授

Keiichi Omoto
— **Professor emeritus of the University of Tokyo, a leading anthropologist and lepidopterist**

尾本惠市（1933～）

尾本惠市博士（1933年～）は東京都出身。1952年東京大学に入学、1972年に東京大学で理学博士号を取得。1979年から東京大学理学部教授となり、1993年から1999年まで国際日本文化研究センター教授を務め、その後は桃山学院大学教授などを歴任した。東京大学および国際日本文化研究センターの名誉教授でもある。専門の分子人類学であるが、チョウ類の研究者としても著名で、特にウスバシロチョウ亜科の系統分類学的研究で大きな成果を挙げている。また、1961年から1964年までドイツに留学し、滞在中の1963年にアフガニスタンのヒンドゥークシ山脈で2ヶ月半ほどチョウ類調査を行い、アウトクラトールウスバシロチョウやマルコポーロモンキチョウのような珍種を自身の手で採集している。将棋に対する造詣の深さでも知られる。

尾本コレクションの標本総数は約27,000点で、マルコポーロモンキチョウの亜種クシャナをはじめ、新種・新亜種の記載に用いられたホロタイプ標本9点を含む多くのタイプ標本が収納されている。収集の重点は、ユーラシアやアメリカ北部、オセアニアに産するアゲハチョウ科のウスバアゲハ亜科、シロチョウ科のミヤマシロチョウ属・モンキチョウ属、タテハチョウ科のコムラサキ亜科、ジャノメチョウ亜科などの研究用標本である。本コレクションは2013年に東京大学総合研究博物館へ寄贈された。

（矢後勝也）

Dr. Keiichi Omoto (1933–), who was born in Tokyo, is a famous lepidopterist who has dedicated himself to researching butterflies. He began a period of academia at the University of Tokyo in 1952, receiving a doctorate of science in 1972. Subsequently, he became a professor of the Faculty of Science, the University of Tokyo, going on to work as a professor at the International Research Center for Japanese Studies in 1993–99, then as a professor at Momoyama Gakuin University. He is currently a professor emeritus of the University of Tokyo and the International Research Center for Japanese Studies. His primary field is molecular anthropology, but he is also prominent as an entomologist, studying the systematics of butterflies, primarily Parnassiinae of the family Papilionidae. In 1963, while studying in Germany between 1961 and 1964, he carried out butterfly research with Colin W. Wyatt at the Hindu Kush mountains, Afghanistan, collecting very rare species such as *Parnassius autocrator* and *Colias marcopolo*. He is also known to be an accomplished player of shogi, Japanese chess.

His butterfly collection, which is comprised of roughly 27,000 specimens, contains many type specimens including nine holotypes originally described by himself such as the subspecies *kushana* of *Colias marcopolo*. Of particular interest are Parnassiinae of the Papilionidae, *Apporia* and *Colias* of the Pieridae, and Apaturinae and Satyrinae of the Nymphalidae from Eurasia, North America and Oceania. In 2013, this collection was kindly donated to the University Museum, the University of Tokyo.

(Masaya Yago)

1

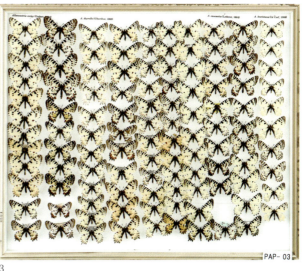

3

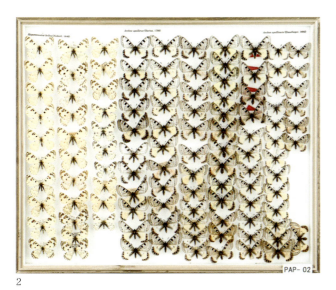

2

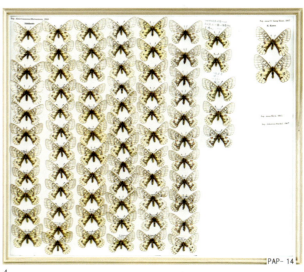

4

1. ウラギンアゲハ・ウスバジャコウアゲハ他 [アゲハチョウ科] (鱗翅目).
   Balonia brevicornis, Cressida cressida, etc. [Papilionidae] (Lepidoptera).

2. イランアゲハ・シリアアゲハ [アゲハチョウ科] (鱗翅目).
   Hypermnestra helios, Archon apollinus [Papilionidae] (Lepidoptera).

3. シロタイスアゲハ・デイロールシロタイスアゲハ他 [アゲハチョウ科] (鱗翅目).
   Allancastria cerisyi, A. deyrollei, etc. [Papilionidae] (Lepidoptera).

4. ウスバキチョウ [アゲハチョウ科] (鱗翅目).
   Parnassius eversmanni [Papilionidae] (Lepidoptera).

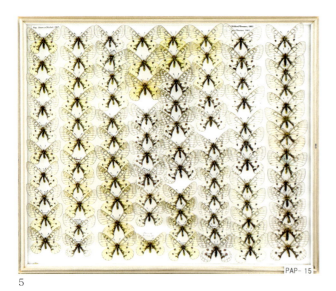

5

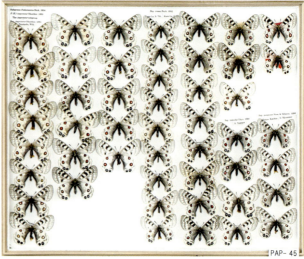

7

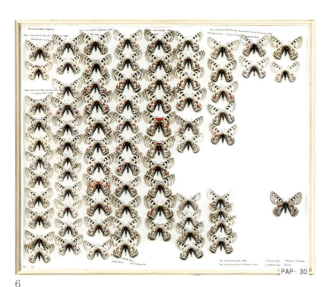

6

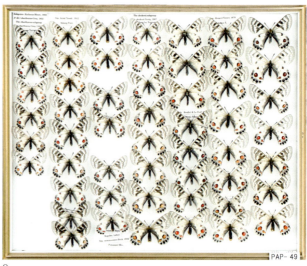

8

5. ウスバキチョウ［アゲハチョウ科］（鱗翅目）.
   Parnassius eversmanni [Papilionidae] (Lepidoptera).

6. アッコウスバシロチョウ［アゲハチョウ科］（鱗翅目）.
   Parnassius acco [Papilionidae] (Lepidoptera).

7. ミカドウスバシロチョウ［アゲハチョウ科］（鱗翅目）.
   Parnassius imperator [Papilionidae] (Lepidoptera).

8. チャールトンウスバシロチョウ［アゲハチョウ科］（鱗翅目）.
   Parnassius charltonius [Papilionidae] (Lepidoptera).

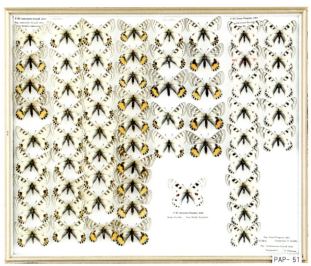

9

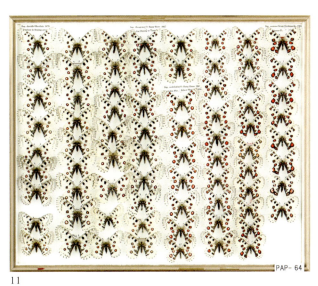

11

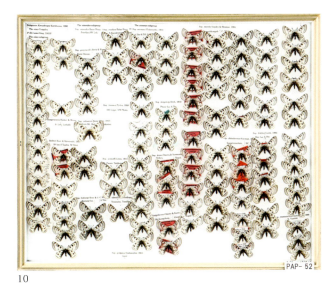

10

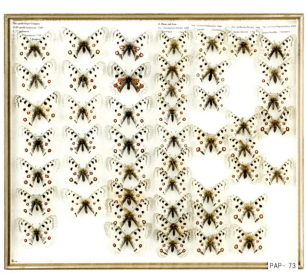

12

9. アウトクラトールウスバシロチョウ・ダヴィドフウスバシロチョウ・ロキシアス
   ウスバシロチョウ［アゲハチョウ科］(鱗翅目).
   Parnassius autocrator, P. davydovi, P. loxias [Papilionidae] (Lepidoptera).

10. シモウスバシロチョウ［アゲハチョウ科］(鱗翅目).
    Parnassius simo [Papilionidae] (Lepidoptera).

11. オオアカボシウスバシロチョウ［アゲハチョウ科］(鱗翅目).
    Parnassius nomion [Papilionidae] (Lepidoptera).

12. アポロウスバシロチョウ［アゲハチョウ科］(鱗翅目).
    Parnassius apollo [Papilionidae] (Lepidoptera).

13

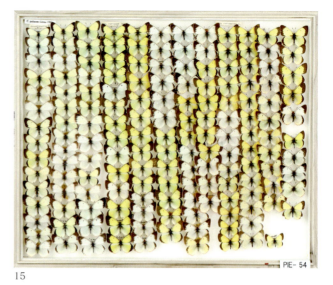

15

14

16

13. エスペランサトラフアゲハ・ガラマスアゲハ［アゲハチョウ科］他（鱗翅目）.
    Pterourus esperanza, P. garamas [Papilionidae], etc. (Lepidoptera).

14. メスアカモンキアゲハ・ホシボシアゲハ［アゲハチョウ科］他（鱗翅目）.
    Pterourus esperanza, P. garamas [Papilionidae], etc. (Lepidoptera).

15. ミヤマモンキチョウ［シロチョウ科］（鱗翅目）.
    Colias palaeno [Pieridae] (Lepidoptera).

16. カナダモンキチョウ・ツンドラモンキチョウ［シロチョウ科］（鱗翅目）.
    Colias canadensis, C. hecla [Pieridae] (Lepidoptera).

17

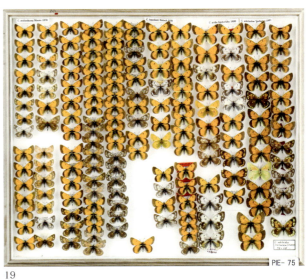

19

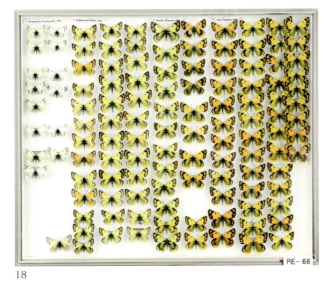

18

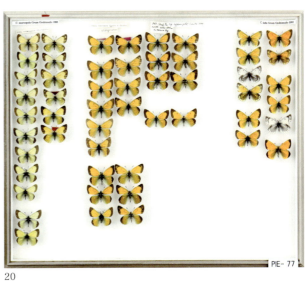

20

17. アムールモンキチョウ［シロチョウ科］（鱗翅目）.
    Colias tyche [Pieridae] (Lepidoptera).

18. ラダックモンキチョウ・ニナモンキチョウ他［シロチョウ科］（鱗翅目）.
    Colias ladakensis, *C. nina*, etc. [Pieridae] (Lepidoptera).

19. ストリカーナモンキチョウ・アデライダモンキチョウ他［シロチョウ科］（鱗翅目）.
    Colias stoliczkana, *C. adelaidae*, etc. [Pieridae] (Lepidoptera).

20. マルコポーロモンキチョウ・ラダモンキチョウ［シロチョウ科］（鱗翅目）.
    Colias marcopolo, *C. lada* [Pieridae] (Lepidoptera).

21

23

22

24

136

21. ウィスコットモンキチョウ［シロチョウ科］（鱗翅目）．
    Colias wiskotti [Pieridae] (Lepidoptera).

22. ムモンアカシジミ・チョウセンアカシジミ・ウラクロシジミ他［シジミチョウ科］
    （鱗翅目）．
    Shirozua jonasi, Coreana raphaelis, Iratsume orsedice, etc. [Lycaenidae]
    (Lepidoptera).

23. フジミドリシジミ・キリシマミドリシジミ・ヒサマツミドリシジミ他［シジミチョウ科］
    （鱗翅目）．
    Sibataniozephyrus fujisanus, Thermozephyrus ataxus, Chrysozephyrus hisamatsusanus,
    etc. [Lycaenidae] (Lepidoptera).

24. アサマシジミ・オオルリシジミ・ゴマシジミ他［シジミチョウ科］（鱗翅目）．
    Plebejus subsolanus, Shijimiaeoides divinus, Phengaris teleius, etc. [Lycaenidae]
    (Lepidoptera).

25

27

26

28

25. チョウセンメスアカシジミ・フチグロベニシジミ・アリオンゴマシジミ他 ［シジミチョウ科］（鱗翅目）．
Thecla betulae, Heodes virgaureae, Phengaris arion, etc. [Lycaenidae] (Lepidoptera).

26. ルリアリヒスイシジミ・ゲノベヴァヤドリギシジミ・デリキアニシキシジミ他 ［シジミチョウ科］（鱗翅目）．
Jalmenus evagoras, Ogyris genoveva, Hypochrysopus delicia, etc. [Lycaenidae] (Lepidoptera).

27. アンビカコムラサキ・ミスジコムラサキ他［タテハチョウ科］（鱗翅目）．
Mymathyma ambica, M. chevana, etc. [Nymphalidae] (Lepidoptera).

28. ゴマダラチョウ［タテハチョウ科］（鱗翅目）．
Hestina persimilis [Nymphalidae] (Lepidoptera).

29

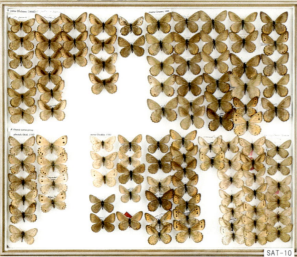
30

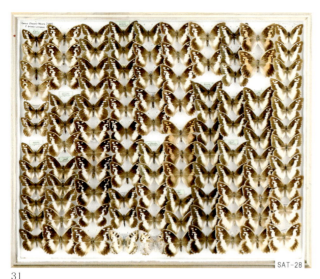
31

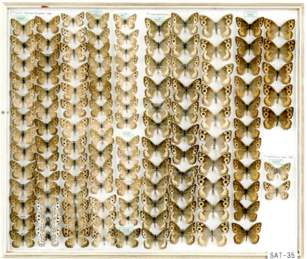
32

33

29. クロオオムラサキ［タテハチョウ科］（鱗翅目）.
    Sasakia funebris [Nymphalidae] (Lepidoptera).

30. ユッタタカネヒカゲ・マグナタカネヒカゲ・タカネヒカゲ［タテハチョウ科］（鱗翅目）.
    Oenies jutta, O. magna & *O. norna* [Nymphalidae] (Lepidoptera).

31. ブリセイスシロオビジャノメ［タテハチョウ科］（鱗翅目）.
    Chazara briseis [Nymphalidae] (Lepidoptera).

32. ブリセイスシロオビジャノメ［タテハチョウ科］（鱗翅目）.
    Pseudochazara briseis [Nymphalidae] (Lepidoptera).

33. ゲイエールニセシロオビジャノメ・キンゴフスキーニセシロオビジャノメ・ダゲスタナニセシロオビジャノメ［タテハチョウ科］（鱗翅目）.
    Pseudochazara geyeri, P. cingovskii & *P. daghestana* [Nymphalidae] (Lepidoptera).

1-33. 箱寸法 (Box dimantions) 418×507 mm.

岸田泰則
——日本蛾類学会会長

Yasunori Kishida
— President of the Japan Heterocerists' Society

岸田泰則（1949〜）

岸田泰則氏（1949年〜）は、世界的に著名な蛾類研究者として知られる。東京都世田谷区北沢で生まれ、幼少からの昆虫好きが高じて中学より本格的に昆虫を収集した。東京農業大学附属第一高等学校に入学してからは主に蛾類の研究を始め、1967年に東京農業大学に進学し、昆虫学研究室に所属しながら蛾類の研究を続けた。1971年に大学卒業後、1973年から田園調布雙葉学園、1976年から宝仙学園中・高等学校の理科教諭として勤務し、2008年に早期退職しているが、この間も国内外問わず精力的に採集に赴きながら、ヒトリガ科を主軸とした大蛾類の分類学研究に没頭し、200を超える多数の論文、報文を発表している。「日本産蛾類標準図鑑 全4巻（学研）」、「日本産幼虫図鑑（学研）」、「エンドレス・コレクションシリーズ Vol. 8 昼蛾（ESI）」、「世界の美麗ヒトリガ（むし社）」などに代表される著書や編書の出版も多く、蛾類分野を中心に昆虫分類学の発展に大きく寄与している。その一方で、2010年から2012年の3年間は日本鱗翅学会会長を歴任した他、2001年から2018年現在まで長く日本蛾類学会会長も務めている。

岸田昆虫コレクションは主にガ類からなり、一部にチョウ類やカミキリ類、セミ類など多岐におよぶ。今日でも岸田氏は蛾類を中心に多岐に渡った昆虫の収集、研究を精力的に続けており、そのコレクションも充実の一途を辿っているが、岸田氏ご自身により採集されたものをはじめ、交流のある昆虫収集家諸氏のご協力により集められた標本も多い。今でもガ類の新種を記載し続けており、それらのタイプ標本も東京大学総合研究博物館に絶えず蓄積されつつある。

（矢後勝也）

Mr. Yasunori Kishida (1949 –) is known as an esteemed taxonomist specializing in moths. Born in Setagaya-ku, Tokyo, he first became fascinated with insects in his childhood, becoming properly absorbed in collecting them during his junior high school days. Subsequently, he entered the Tokyo University of Agriculture Dai-ichi High School, where he began to primarily research moths. He then went on to Tokyo University of Agriculture in 1967, where he conducted taxonomic studies of mainly arctiid moths in the Laboratory of Entomology until his graduation in 1971. He later taught science at the Denenchofufutaba Gakuen Junior and Senior High School from 1973, then at the Hosen Gakuen Junior and Senior High School from 1976, before finally taking early retirement in 2008. During this time, he energetically collected moths and other insects both in Japan and overseas, in addition to publishing more than 200 publications on taxonomic studies of macrolepidopterans including the widely-regarded titles "The Standard of Moths in Japan, vol. 1-4 (Gakken)", "Insect Larvae of Japan (Gakken)", "Endless Collection Series Vol. 8. Day-Flying Moths (ESI)" and "Handbook Series of Insects 5: Selected Arctiid Moths of the World (Mushi-sha)". His work has contributed to the development of insect taxonomy, especially in the area of Lepidoptera. Moreover, he has also held prominent posts such as President of the Lepidopterological Society of Japan (2010-2012) and is currently President of the Japan Heterocerists' Society (2001-present).

The Kishida insect collection, which comprises roughly 500 specimen boxes, is composed mainly of moths, with the remainder being butterflies, longicorn beetles, cicadas and others. The majority of the specimens were captured himself, and the remaining specimens by his collaborators. He continues to passionately research and collect various insects, especially moths, with his insect collection growing steadily. He also continues to describe new species of moths, of which the type specimens are constantly being stored in the University Museum, the University of Tokyo.

(Masaya Yago)

3

1

4

2

140

1. キボシルリニシキ・ベニモンマダラ［マダラガ科］・クビアカスカシバ［スカシバガ科］他（鱗翅目）.
Amesia sanguiflua & *Zygaena niphona* [Zygaenidae], *Glossophecia romanovi* [Sesiidae], etc. (Lepidoptera).

2. マエベニノメイガ・クロスジノメイガ・シロフクロノメイガ他［ツトガ科］（鱗翅目）.
Paliga minnehaha, Tyspanodes striatus, Pygospila tyres, etc. [Crambidae] (Lepidoptera).

3. ヒロバカレハ・ウスズミカレハ・スカシカレハ他［カレハガ科］（鱗翅目）.
Gastropacha quercifolia, Poecilocampa takamukui, Amurilla subpurpurea, etc. [Lasiocampidae] (Lepidoptera).

4. マツカレハ・イワサキカレハ・ウスマダラカレハ他［カレハガ科］（鱗翅目）.
Dendrolimus spectabilis, Kunugia iwasakii, Pyrosis idiota, etc. [Lasiocampidae] (Lepidoptera).

7

5

8

6

5. クロウスタビガ［ヤママユガ科］・イボタガ［イボタガ科］・クワコ［カイコガ科］他（鱗翅目）.
   Rhodinia jankowskii [Saturniidae], *Brahmaea japonica* [Brahmaeidae], *Bombyx mandarina* [Bombycidae], etc. (Lepidoptera).

6. カエサルサン・ヘラクレスサン［ヤママユガ科］（鱗翅目）.
   Attacus caesar & *Coscinocera hercules* [Saturniidae] (Lepidoptera).

7. ヒメベニオナガミズアオ・ボルネオオオヤママユ他［ヤママユガ科］（鱗翅目）.
   Actias dubernardi, *Archaeoattacus staudingeri*, etc. [Saturniidae] (Lepidoptera).

8. オナガヤママユ・マダガスカルオナガヤママユ他［ヤママユガ科］（鱗翅目）.
   Actias maenas, *Argema mittrei*, etc. [Saturniidae] (Lepidoptera).

11

9

12

10

142

9. オオミズアオ・オナガミズアオ他［ヤママユガ科］（鱗翅目）.
   Actias ariena, A. gnoma, etc. [Saturniidae] (Lepidoptera).

10. クロメンガタスズメ・ヨーロッパメンガタスズメ・オオシモフリスズメ他［スズメガ科］（鱗翅目）.
    Acherontia lachesis, A. atropos, Langia zenzeroides, etc. [Sphingidae] (Lepidoptera).

11. ケブカスズメ・イブキスズメ・フリッツェホウジャク他［スズメガ科］（鱗翅目）.
    Pentateucha stuenigi, Hyles gallii, Macroglossum frizei, etc. [Sphingidae] (Lepidoptera).

12. カラフトアヤトガリバ・モントガリバ・キボシミスジトガリバ他［カギバガ科］（鱗翅目）.
    Habrosyne intermedia, Thyatira batis, Achlya longipennis, etc. [Drepanidae] (Lepidoptera).

13

15

16

14

13. クロフシロナミシャク・モンクロキイロナミシャク・キマダラオオナミシャク他［シャクガ科］（鱗翅目）．
Otoplecta frigida, Stamnodes danilovi, Gandaritis fixseni, etc. [Geometridae] (Lepidoptera).

14. オガサワラチズモンアオシャク［シャクガ科］・オガサワライラガ［イラガ科］・タカネヨトウ［ヤガ科］他（鱗翅目）．
Agathia ichnospora [Geometridae], *Belippa boninensis* [Limacodidae], *Sympistis heliophile* [Noctuidae], etc. (Lepidoptera).

15. クロツマキシャチホコ・モンクロシャチホコ他［シャチホコガ科］（鱗翅目）．
Phalera minor, P. flavescens, etc. [Notodontidae] (Lepidoptera).

16. ウチキシャチホコ・マエジロシャチホコ他［シャチホコガ科］（鱗翅目）．
Notodonta dembowskii, N. albicosta, etc. [Notodontidae] (Lepidoptera).

19

17

20

18

17. モンクロギンシャチホコ・アカシャチホコ・モンキシロシャチホコ他［シャチホコガ科］（鱗翅目）.
    Wilemanus bidentatus, Gangaridopsis citrina, Leucodonta bicoloria, etc. [Notodontidae] (Lepidoptera).

18. カバイロシャチホコ・ウスキシャチホコ・キンイロシャチホコ他［シャチホコガ科］（鱗翅目）.
    Ramesa tosta, Mimopydna pallida, Eushachia aurata, etc. [Notodontidae] (Lepidoptera).

19. ギンモンスズメモドキ・タイワンギンモンスズメモドキ他［シャチホコガ科］（鱗翅目）.
    Tarsolepis japonica, T. taiwana, etc. [Notodontidae] (Lepidoptera).

20. コシロオビドクガ・スキバドクガ・トラサンドクガ他［ドクガ科］（鱗翅目）.
    Numenes disparilis, Perina nuda, Kidokuga torasan, etc. [Lymantriidae] (Lepidoptera).

21

23

22

24

21. ベニモンコノハ・ナンベイオオヤガ［ヤガ科］・カシミールウズマキヒトリ［ヒトリガ科］他（鱗翅目）.
    Phyllodes consobrinus, *Thysania agrippina* [Noctuidae], *Gonerda perornata* [Arctiidae], etc. (Lepidoptera).

22. アオスジアオリンガ・ウスベニアオリンガ・サラサリンガ他［コブガ科］（鱗翅目）.
    Pseudoips prasinanus, *Earias erubescens*, *Camptoloma interioratum*, etc. [Nolidae] (Lepidoptera).

23. キシタアオバケンモン・ミツモンケンモン・エゾサクラケンモン他［ヤガ科］（鱗翅目）.
    Euromoia subpulchra, *Cymatophoropsis trimaculata*, *Acronicta strigosa*, etc. [Noctuidae] (Lepidoptera).

24. キシタミドリヤガ・オオアオバヤガ他［ヤガ科］（鱗翅目）.
    Xestia efflorescens, *Anaplectoides virens*, etc. [Noctuidae] (Lepidoptera).

25

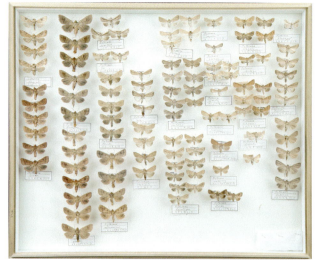

27

26

28

25. アオアカガネヨトウ・コゴマヨトウ・クロビロードヨトウ他 [ヤガ科] (鱗翅目).
    Karana laetevirens, Chandata bella, Sidemia bremeri, etc. [Noctuidae] (Lepidoptera).

26. アフリカシロナヨトウ・オスキバネヨトウ・ウグイスセダカヨトウ他 [ヤガ科] (鱗翅目).
    Spodoptera exempta, Athetis thoracica, Mormo cyanea, etc. [Noctuidae] (Lepidoptera).

27. ノコバヨトウ・ミヤマフタオビキヨトウ・カバイロキヨトウ他 [ヤガ科] (鱗翅目).
    Tiracola plagiata, Mythimna matsumuriana, M. iodochra, etc. [Noctuidae] (Lepidoptera).

28. オオシモフリヨトウ・シラホシヨトウ・ハマオモトヨトウ他 [ヤガ科] (鱗翅目).
    Polia goliath, Melanchra persicariae, Brithys crini, etc. [Noctuidae] (Lepidoptera).

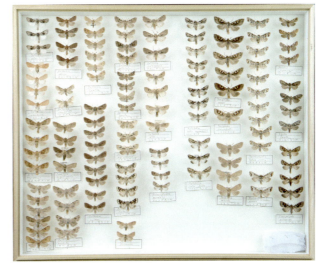
31

29

32

30

29. イチモジヒメヨトウ・エゾショウブヨトウ他[ヤガ科]（鱗翅目）.
　　Xylomoia fusei, Amphipoea lucens, etc. [Noctuidae] (Lepidoptera).

30. オオハガタヨトウ・キマエキリガ・スギタニモンキリガ他[ヤガ科]（鱗翅目）.
　　Mniotype melanodonta, Hemiglaea costalis, Sugitania lepida, etc. [Noctuidae] (Lepidoptera).

31. ナマリキリガ・タカオキリガ・アズサキリガ他[ヤガ科]（鱗翅目）.
　　Orthosia satoi, Pseudopanolis takao, P. azusa, etc. [Noctuidae] (Lepidoptera).

32. シロスジキリガ・キバラモクメキリガ・ウスアオキリガ他[ヤガ科]（鱗翅目）.
　　Lithomoia solidaginis, Xylena formosa, Lithophane venusta, etc. [Noctuidae] (Lepidoptera).

35

33

36

34

33. タニガワモクメキリガ・タカセモクメキリガ・ケンモンミドリキリガ他[ヤガ科]（鱗翅目）．
    Brachionycha permixta, B. albicilia, Daseochaeta viridis, etc. [Noctuidae] (Lepidoptera).

34. アオモンギンセダカモクメ・トラガ・マイコトラガ他[ヤガ科]（鱗翅目）．
    Cucullia argentea, Chelonomorpha japana, Maikona jezoensis, etc. [Noctuidae] (Lepidoptera).

35. クロシタホソヤガ・ソリバネホソヤガ・オビナカジロフサヤガ他[ヤガ科]（鱗翅目）．
    Odontodes uniformis, Aegilia describens, Penicillaria maculata, etc. [Noctuidae] (Lepidoptera).

36. タンポキンウワバ・エゾヒサゴキンウワバ・アルプスギンウワバ他[ヤガ科]（鱗翅目）．
    Autographa excelsa, Diachrysia chrysitis, Syngrapha ottolenguii, etc. [Noctuidae] (Lepidoptera).

39

37

40

38

37. ベニトガリアツバ・ナカオビシロコヤガ・ミジンベニコヤガ他［ヤガ科］（鱗翅目）.
    Naganoella timandra, Eublemma rivula, Ectoblemma rosella, etc. [Noctuidae] (Lepidoptera).

38. ムラサキシタバ・シロシタバ・アズミキシタバ他［ヤガ科］（鱗翅目）.
    Catocala fraxini, C. nivea, C. koreana, etc. [Noctuidae] (Lepidoptera).

39. カバフキリバ・シラフクチバ・ニジオビベニアツバ他［ヤガ科］（鱗翅目）.
    Episparis okinawensis, Sypnoides picta, Homodes vivida, etc. [Noctuidae] (Lepidoptera).

40. クロモンハイイロクチバ・スソミダレアツバ・キシダクビグロクチバ他［ヤガ科］（鱗翅目）.
    Bamra albicola, Pilipectus prunifera, Lygephila kishidai, etc. [Noctuidae] (Lepidoptera).

1-40. 箱寸法 (Box dimantions) 418×507 mm.

第3章

Chapter 3

幻の大蝶「ブータンシボリアゲハ」
—— ブータン国王陛下からの贈呈標本

Ludlow's Bhutan Glory, a mysterious swallowtail
— Butterfly specimen presented by His Majesty the King of Bhutan

1. ブータンシボリアゲハ♂. ブータン王国タシヤンツェ県トブラン, 2011年8月13日, 矢後勝也(東京大学総合研究博物館)採集.
Bhutanitis ludlowi Gabriel, 1942. Male, 13. Aug. 2011, Tobrang, Trashiyangtse, Bhutan. Dr. Masaya Yago (The University Museum, The University of Tokyo) leg.

ブータンシボリアゲハ *Bhutanitis ludlowi* Gabriel, 1942 は、1933年と1934年の8月にイギリスのプラントハンター Frank Ludlow と George Sherriff によりブータン北東部で発見された。その後、これまで多くの研究者がこの蝶の再発見に挑んできたが、追加の記録は一切途絶えていた。

ブータンのウゲン・ワンチュク保全環境研究所の職員 Karma Wangdi は、2010年8月にブータン北東部のタシヤンツェ渓谷を訪れ、2頭のブータンシボリアゲハと思われる個体を撮影した。この情報をきっかけに、日本蝶類学会のメンバー数名はブータン政府関係者らに許可申請の交渉を重ね、翌年の夏、農林省と日本調査隊との共同学術調査隊を結成し、調査の特別許可が交付された。

2011年8月、共同学術調査隊はタシヤンツェ渓谷でチョウ類調査を行い、最初の発見以来、実に約80年ぶりとなる本種を3♂2♀採集することに成功した。また、同時に65〜180卵からなる山積みの卵塊を産卵するという本種の特異な生態も明らかにされたのである。一方で、ワシントン条約等の厳しい制限のために、輸出申請は行ってきたものの日本への持ち出し許可がその場では得られず、本種の形態学的研究や分子系統学的研究が難しい状況であった。ところが、同年11月に国賓で日本を訪れたブータン国王陛下夫妻が、国家間の友好の証と東日本大震災からの復興の願いを込めて、豪華な標本箱に収納された2♂のブータンシボリ標本をお運び下さり、日本調査隊隊員が所属する東京大学総合研究博物館と進化生物学研究所にそれぞれ贈呈された。今回展示された標本は本特別展の企画・総指揮を行った編著者の矢後による採集個体である。これにより、シボリアゲハ属4種間での形態比較や分子系統の研究が可能となった。

(矢後勝也)

Ludlow's Bhutan Glory, *Bhutanitis ludlowi* Gabriel, 1942, was discovered in northeastern Bhutan in Aug. 1933 and 1934 by two English botanical collectors, Frank Ludlow and George Sherriff. Since its discovery many researchers and collectors have attempted without success to rediscover this mysterious swallowtail.

In Aug. 2009, Mr. Karma Wangdi, working at the Ugyen Wangchuk Institute for Conservation and Environment, succeeded in photographing what was potentially two adults of the species at Trashiyangtse Valley, northeastern Bhutan. Based on this, members of the Butterfly Society of Japan (BSJ) submitted a research proposal to study B. ludlowi in this area. After six months of negotiation, special permission was given to undertake an investigation of the Bhutanese butterfly fauna by a joint research team between the Ministry of Agriculture and Forests, Bhutan (MoAF) and BSJ.

In Aug. 2011, the joint research team surveyed the Trashiyangtse Valley where they successfully rediscovered Ludlow's Bhutan Glory after a period of about 80 years. Three males and two females were collected in accordance with the permitted limit. At the same time, they also revealed a unique biology, namely that this butterfly lays its eggs in a mound composed of 65-180 eggs. However, due to the strict export embargo created by the Convention on International

Trade in Endangered Species of Wild Fauna and Flora (CITES), morphological and molecular studies of the butterfly were not possible at that time. In Nov. 2011, His Majesty the King of Bhutan and Her Majesty the Queen visited Japan as state guests. On that great occasion, two ornately-mounted specimens of *B. ludlowi* were gifted to two institutes to which three members of the joint research team belong, as part of ongoing friendly relations between the countries, and as a token of hope for recovery from the Great East Japan Earthquake. This exhibited specimen was collected by Dr. Yago, one of the authors and the Planner and Executive Producer of the special exhibition. These invaluable specimens enabled us to undertake studies among all four *Bhutanitis* species.

(Masaya Yago)

2. Exterior box covered with embroidery of the Royal Crest and paired dragons indicating the symbol of Bhutan.

3. Specimen box with the Royal Crest and paired dragons as in exterior box. A massage "A Gift from the People of Bhutan" are inscribed on the lower side.

4. Top: Name card of His Majesty The King of Bhutan. Middle: Massage card written as "With the Compliments of His Majesty The King of Bhutan". Bottom: Envelop in which the massage card was enclosed.

5

6

7

8

5. ブータン・日本 共同学術調査隊のメンバー.
   Members of Bhutan-Japan joint research team.

6. ブータン東部タシヤンツェ渓谷でのブータンシボリアゲハ調査(2011年8月13日).
   Field research of *Bhutanitis ludlowi* in Trashiyangtse Valley, eastern Bhutan (13. Aug. 2011).

7. ブータンシボリアゲハ♀. ブータン王国タシヤンツェ県トブラン, 2011年8月18日.
   Bhutanitis ludlowi Gabriel, 1942. Female, 18. Aug. 2011, Tobrang, Trashiyangtse.

8. ブータンシボリアゲハの交尾(下:♂). ブータン王国タシヤンツェ県トブラン, 2011年8月16日.
   Copulation of *Bhutanitis ludlowi* (bottom: male). 16. Aug. 2011, Tobrang, Trashiyangtse.

9

10

11

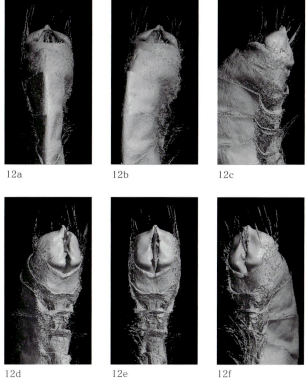

12a　12b　12c

12d　12e　12f

9. ブータンシボリアゲハの産卵（食草グリフィスウマノスズクサ）．ブータン王国タシヤンツェ県トブラン, 2011年8月18日．
 Oviposition of *Bhutanitis ludlowi* (Hostplant: Aristolochia griffithii). 18. Aug. 2011, Tobrang, Trashiyangtse.

10. ブータンシボリアゲハの卵塊．ブータン王国タシヤンツェ県トブラン, 2011年8月18日．
 Egg cluster of *Bhutanitis ludlowi*. 18. Aug. 2011. Tobrang, Trashiyangtse.

11. 赤坂迎賓館でのブータンシボリアゲハ標本の贈呈式（2011年11月15日）．最左が本書編著者の矢後．
 Presentation ceremony of *Bhutanitis ludlowi* specimens at the State Guest House, Akasaka Palace, Tokyo (Akasaka Geihinkan) on 15 Nov. 2011. The author and editor of this book, Dr. Masaya Yago, on the left.

12. ブータンシボリアゲハ♂交尾器のマイクロCT画像（前川　優撮影）．
 Micro-CT Images of male genitalia in *Bhutanitis ludlowi* (Photos by Yu Maekawa).

昆虫
――東京大学総合研究博物館データベース（UMDB）

INSECTS
― The University Museum Database, the University of Tokyo (UMDB)

現在、東京大学総合研究博物館では、収蔵されているコレクションのデータベースを作成して、出版物やウェブ上に公開発信する計画（UMDB プロジェクト）が進められている。本展示では、すでに当館の標本・資料報告にて出版、ホームページ上で公開されている昆虫標本のデータベースをこのコーナーで見られるようにした。現在、佐々木忠次郎教授関連コレクション、岸田泰則コレクション、加藤正世コレクション、老田正夫コレクション、白石浩次郎コレクション、江田　茂コレクション、五十嵐邁コレクション、須田孫七コレクションのカタログが公開されている。このようなデータベースには、分類学や体系学、形態学のような分野に多大な貢献が見込まれるだけでなく、生物多様性保全の基礎となるインベントリー（目録）作成としても捉えられる。分布情報を活用することで、生物地理学の他、地球温暖化や森林破壊のような環境問題を考える上での基礎的情報も多く提供されうる。本データベース公開により、国内外の様々な分野の研究に寄与し、学術標本等の重要性を広く認知させるものとなるだろう。

（矢後勝也）

1. UMDB "昆虫" の展示.
   Exhibition of UMDB "INSECTS".

　The University Museum, The University of Tokyo, is currently data-basing the deposited collections to make publicly available as the University Museum Database (UMDB) project. In this special exhibition, you can browse the database of insect specimens that have been published in the Material Reports series and on the website of the museum. At present, the following catalogues are available: The Insect Collection of Prof. Chûjirô Sasaki and associated researchers, the Yasunori Kishida Insect Collection, the Masayo Kato Insect Collection, the Masayo Kato Fungi Collection, the Suguru Igarashi Insect Collection, the Shigeru Eda Insect Collection, the Kojiro Shiraishi Insect Collection, the Masao Oita Insect Collection and the Magoshichi Suda Insect Collection. This database not only contributes much to taxonomy and systematics, but also provides an inventory that underpins biodiversity conservation. For example, basic data on distribution is the foundation for spatial analysis of patterns in biogeography and changes in geographic range that may be related to global warming and deforestation. We hope that by publishing this database, it will not only contribute to various scientific fields of endeavor, but also promote the importance of preserving and cataloging scientific specimens and creating museum collections to the public.

(Masaya Yago)

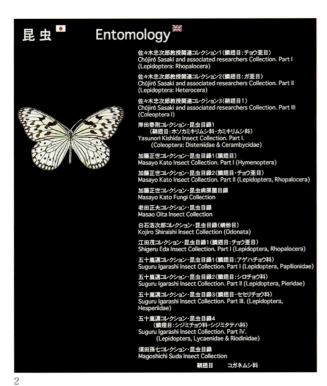

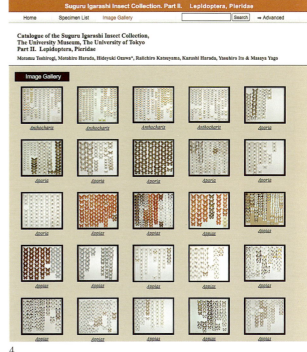

2. UMDB "昆虫" のトップページ.
   Top Page of UMDB "INSECTS".

3. UMDB "昆虫" 内にある「五十嵐邁昆虫コレクション目録（鱗翅目：シロチョウ科）」のトップページ.
   Top Page of "Catalogue of the Suguru Igarashi Insect Collection (Lepidoptera: Pieridae)" in UMDB "INSECTS".

4. 「五十嵐邁昆虫コレクション目録（鱗翅目：シロチョウ科）」内にある画像ギャラリー.
   Image Gallery of specimen boxes in "Catalogue of the Suguru Igarashi Insect Collection (Lepidoptera: Pieridae)".

5. 「五十嵐邁昆虫コレクション目録（鱗翅目：シロチョウ科）」内にある標本リストの一例.
   An example of the Specimen List in "Catalogue of the Suguru Igarashi Insect Collection (Lepidoptera: Pieridae)".

未来に向けて

Looking Towards the Future

最後に、ここで縦一列に展示された空の標本箱は、今後築かれる人と自然との未来を示したものである。近年の環境破壊や地球温暖化などの影響を大きく受けて、昆虫達が激減している今、環境指標となる昆虫標本がこれらの標本箱の中に果たしてどのくらい集積できるのか？ あるいは標本箱に収蔵されるような昆虫が将来的にいなくなるほど環境が激変してしまうのか？ 来館者の方々にも、今後の未来に向けて、生態系への影響を緩和できる手段を考えて頂くきっかけになれば、という想いを込めた。標本箱を空のままであり続けないような未来を創り、そして子供や孫たちに貴重な自然や昆虫を残していくことは、今後の我々人類の存続にとっても極めて重要なことであろう。

(矢後勝也)

Shown here is a visual representation of the future relationship of mankind and nature, indicated by the column of empty specimen boxes. Recently, the number of insects is drastically decreasing due to anthropogenic impacts such as environmental destruction and global warming. In this situation, will insect specimens, effective as environmental indicators, be accumulated in these specimen boxes in the future or will the environment change dramatically enough that there are no insects to be stored in the boxes? We feel that this is an opportunity for visitors to think about the future and what we can do to ease our impacts on the ecology. To build a future that will not result in these boxes remaining empty, a future which maintains the natural world and the insects in it could be vitally important, not only for our children and grandchildren, but for the survival of mankind in the future.

(Masaya Yago)

第4章

Chapter 4

展示制作
——立案から完成まで

Exhibition production
— From plan to completion

今回の特別展では、江戸から明治、大正、昭和、平成と続く昆虫学史に残る昆虫学者達が残した標本とその意義が伝えられる展示を、これまでにない展示手法で試みたいというねらいがあった。また、天井から足元までの壁一面を標本箱で敷き詰められた展示空間の四方から昆虫標本に囲まれることで、各研究者の研究に傾けた情熱や収集に対する執着心のようなものを表現したかった。この項では、その構想の段階から展示が出来上がっていくまでの過程を写真で紹介したい。

(矢後勝也)

The aim of this special exhibition is to raise awareness and recognition of the importance of the insect specimens left by the fifteen entomologists. These specimens provide a visual history of Japanese entomology from the Edo to Heisei eras through the Meiji, Taisho and Showa eras, which we have showcased in an unprecedented manner. Being surrounded by a wall-to-wall, all-encompassing display of the insect specimens, we want to express the enthusiasm and passion of the entomologists and the wonder and excitement of collecting insects. In this section, we introduce the process from the concept phase to the completion of this exhibition using photographs.

(Masaya Yago)

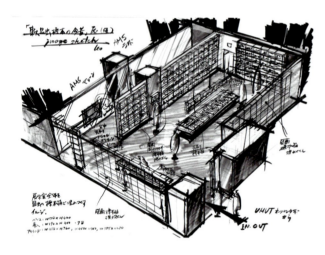

1. 展示場イメージスケッチ.
   Image sketch of exhibition hall.

2. 展示場の模型(俯瞰).
   Scale model of exhibition hall (overhead view).

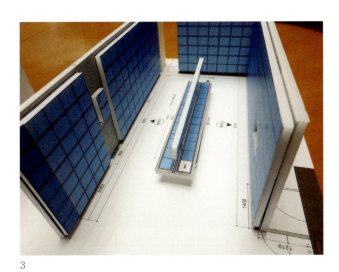

3

5

4

3. 展示場の模型（俯瞰）.
　Scale model of exhibition hall (overhead view).

4. 制作中の展示場（2018年7月10日）.
　Making exhibition (10. Jul. 2018).

5. 完成した展示場（俯瞰）.
　Completed exhibition hall (overhead view).

会場風景
Exhibition views

Photos by Forward Stroke inc.

参考文献

References

Awano, K., Ozawa, H., Yago, M. and Nishino, Y., 2012. Catalogue of the Shigeru Eda Insect Collection, The University Museum, The University of Tokyo. Part I. Lepidoptera, Rhopalocera. *The University Museum, The University of Tokyo, Material Reports,* (93): 1-68.

江崎梯三 , 1953. 日本昆蟲学史話—江戸時代篇 , 9. 新昆蟲 , 6 (1): 20-25.

波部忠重 , 1994. 武蔵石寿 . 下中　弘（編）. 彩色江戸博物学集成 : 209-224. 平凡社 , 東京 .

Harada, M., Karma Wangdi, Sonam Wangdi, Yago, M., Aoki, T., Igarashi, Y., Yamaguchi, S., Watanabe, Y., Sherub, Rinchen Wangdi, Sangay Drukpa, Saito, M., Moriyama, Y. and Uchiyama, T., 2012. Rediscovery of Ludlow's Bhutan Glory, *Bhutanitis ludlowi* Gabriel (Lepidoptera: Papilionidae): morphology and biology. *Butterflies: Journal of the Butterfly Society of Japan,* (60): 4-15.

Harada, M., Teshirogi, T., Ozawa, H. and Yago, M., 2012. Catalogue of the Suguru Igarashi Insect Collection, The University Museum, The University of Tokyo. Part I. Lepidoptera, Papilionidae. *The University Museum, The University of Tokyo, Material Reports,* (94): 1-390.

Harada, M., Teshirogi, M., Ozawa, H., Katsuyama, R., Harada, K., Ito, Y. and Yago, M., 2014. Catalogue of the Suguru Igarashi Insect Collection, The University Museum, The University of Tokyo. Part II. Lepidoptera, Pieridae. *The University Museum, The University of Tokyo, Material Reports,* (99): 1-235.

橋本洽二・林　正美・大野正男 , 1981. 復刻蝉の生物学・別冊資料篇 . サイエンティスト社 , 東京 .

猪又敏男・植村好延・矢後勝也・神保宇嗣・上田恭一郎 , 2013. 日本昆虫目録 第 7 巻 鱗翅目（第 1 号 セセリチョウ上科 － アゲハチョウ上科）. 櫂歌書房 , 福岡 .

Inoue, A., Harada, K., Ito, Y. and Yago, M., 2017. Catalogue of the Yasunori Kishida Insect Collection, The University Museum, The University of Tokyo. Part I. (Coleoptera: Disteniidae and Cerambycidae). *The University Museum, The University of Tokyo, Material Reports,* (112): 1-180.

石森直人 , 1938. 佐々木忠次郎博士を憶ふ . 昆蟲 , 12 (4): 115-120.

鏑木外岐雄 , 1939. 佐々木忠次郎博士記念號追憶編 . 佐々木忠次郎博士 . 動物学雑誌 , 51 (7): 381-390.

加藤正世 , 1933. 温故知新 . 昆蟲界 , 1 (6): 592-601.

小西正泰・田中　誠 , 1985. 虫を針でとめるまで . アニマ , 13 (9): 69-73.

小西正泰 , 1989. 前田利保と黒田斉清—赭鞭会をめぐる人々 . 科学朝日 , 49 (7): 90-94.

小西正泰 , 1993. 虫の博物誌 . 朝日新聞社 , 東京 .

小西正泰 , 1997. 第 1 章 . 混沌の時代—開成所と物産会 . 洋学機関の成立 . 江戸末期と明治前半の昆虫標本—東京大学の所蔵品を中心に . 東京大学（編）, 東京大学創設

百二十周年記念東京大学展―学問の過去・現在・未来，第一部「学問のアルケオロジー」：50-58．東京大学総合研究資料館，東京．

小西正泰，2000．昆虫学のあゆみ―東洋編．インセクタリゥム，37 (11): 10-16．

三宅恒方，1919．昆蟲學汎論，下巻．裳華房，東京．

玉木　存，1998．動物学者箕作佳吉とその時代―明治人は何を考えたか．三一書房，東京．

Nagase, H., Harada, K., Ito, H. and Yago, M., 2014. Catalogue of the Masayo Kato Insect Collection, The University Museum, The University of Tokyo. Part I. Hymenoptera. *The University Museum, The University of Tokyo, Material Reports,* (104): 1-189.

名和　靖，1889．岐阜蝶ノ實驗．動物學雜誌，1 (10): 137-320．

佐々木忠次郎，1926．日本昆蟲學の發達．昆蟲，1 (1): 1-5．

Sato, H., Awano, K. and Yago, M., 2014. Catalogue of the Masayo Kato Fungi Collection, The University Museum, The University of Tokyo. *The University Museum, The University of Tokyo, Material Reports,* (102): 1-69.

白水隆文庫刊行会（編），2007．白水隆アルバム―日本蝶界の回想録―．白水隆文庫刊行会，松江．

新昆虫編集部，1950．寫眞グラフ・寫眞探訪　東京の三大昆虫博物館．新昆虫，3 (4): 1-4（前付）．

Suda, S., Sugiura, Y., Awano, K., Kato, Y., Ito, H., Ito, Y. and Yago, M., 2013. Catalogue of the Kojiro Shiraishi Insect Collection, The University Museum, The University of Tokyo. -Odonata-. *The University Museum, The University of Tokyo, Material Reports,* (98): 1-106.

田付貞洋，1995．25 幕臣武蔵石寿孫右衛門自製昆虫標本．大場秀章・西野嘉章（編），東京大学コレクション (II)―動く大地とその生物：142-144．東京大学総合研究資料館，東京．

著者不明，1913．八十年前に作れる昆蟲標本．昆蟲世界，17: 478．

矢後勝也，2005．日本最古の昆虫標本 (1)-(2) ―東京大学所蔵のチョウ類コレクションから―．昆虫と自然，40 (2): 28-31; 40 (3): 27-31．

矢後勝也，2012．東京大学総合研究博物館案内 241・日本最古の昆虫コレクション ―幕臣・武蔵石寿自製の昆虫標本―．文部科学教育通信，(305): 47．

矢後勝也，2012．生き物のいま．幸福の国に棲む幻の大蝶．ヒマラヤの貴婦人ブータンシボリアゲハ．Biostory, 17: 58-61．

矢後勝也，2012．幻の蝶を追って．文藝春秋，90 (4): 88-90．

Yago, M., Kato, Y., Ito, H., Awano, K., Niitsu, S., Matsubara, H., Ito, Y. and Endo, H., 2014. Catalogue of the Masao Oita Insect Collection, The University Museum, The University

of Tokyo. *The University Museum, The University of Tokyo, Material Reports,* (101): 1-103.

矢後勝也（監修），2015．蟬類博物館―昆虫黄金期を築いた天才・加藤正世博士の世界．練馬区石神井公園ふるさと文化館，東京．

矢後勝也・平井規央・神保宇嗣（編），2016．日本産チョウ類の衰亡と保護 第7集．日本鱗翅学会，東京．

Yoshida, Y., Harada, K., Ito, H., Ito, Y. and Yago, M., 2016. Catalogue of the Masayo Kato Insect Collection, The University Museum, The University of Tokyo. Part II. (Lepidoptera: Rhopalocera). *The University Museum, The University of Tokyo, Material Reports,* (108): 1-238.

Wangdi, K., Harada, M., Yago, M., Wangdi, S., Sherub, Aoki, T., Yamaguchi, S., Igarashi, Y., Watanabe, Y., Wangdi, R., Drukpa, S., Saito, M., Moriyama, Y. and Uchiyama, T., 2012. Expedition report on Ludlow's Bhutan Glory, a mysterious swallowtail butterfly. *Butterflies: Journal of the Butterfly Society of Japan,* (62): 7-15.